许靖华院士简介

许靖华,男,特聘教授,1929年生于南京市,是河南大学老校长许心武教授之子,国际著名的地质学者、海洋学者和环境学者,美国科学院外籍院士(1986)、第三世界科学院外籍院士(1986)、地中海科学院院士(1988)等。因为他对中国地质学的贡献,被授予中国科学院名誉教授(1985)、台湾中央研究院院士(1990)及南京大学荣誉博士等。

学历:1944年考入中央大学(1949年更名南京大学)地质系,1948年7月毕业并获学士学位。同年赴美留学,1950年在俄亥俄州立大学获硕士学位,1953年在洛杉矶加州大学获博士学位。

简历:1954年入美国壳牌石油公司从事研究工作,1963年到1967年在纽约州立大学和加州大学任教。1967年起移居瑞士,任教于瑞士联邦理工大学,任地球科学院院长。曾任国际沉积学会主席、国际海洋地质学委员会主席、国际海洋学委员会执行委员会理事、欧洲地球物理协会首任会长、瑞士联合国教科文组织代表、联合国核废料海底处置专家团专家、联合国的中国政府和马耳他政府技术顾问及原中国地质部、石油地质部和化工部顾问。他被《沉积学报》、《海洋地球物理学研究》等20多家世界级权威学术刊物聘为主编或编委。他担任了许多国际科学合作项目的领导者,如 JOIDES 全球深海钻探项目、上地幔项目、岩石圈项目、全球变化项目以及内蒙古、新疆、西藏的国际考察项目等。

成果与奖励:已经发表 SCI 论文 450 篇,出版中英文著作 20 多部。

他担任过世界各洲的客座教授或客座研究员,是著名的美国古根海姆学者(1971)、伊丽莎白女王海洋地质学高级学者(1981)、Fairchild 教授(1990)等。他是先后获得渥拉斯顿奖(1984)和彭罗斯奖(2001)两个奖的少数科学家之一。渥拉斯顿奖被认为是自然科学界的诺贝尔奖,曾经授予地质学的奠基人威廉·斯密斯、查尔斯·莱尔以及生物进化论的鼻祖查尔斯·达尔文和托马斯·赫胥里。他还获得了其他地学不同领域的最高奖,包括沉积学、沉积地质学、海洋地质学、石油地质学、构造学等。

他还是一位发明家,有多项专利技术。如他发明的"过滤—水转化器"是污水处理的革命性技术,"毛细管灌溉技术"将使缺水的荒地变为绿洲,"石油回采技术"将会缓解石油能源危机。他认识到亚硝酸盐污染和癌症的关系,在实验观察的基础上提出了清晰的论述。因此,他被聘为河南大学环境与健康中心主任。他希望他在学校的工作能够引领发现新方法,解救成千上万的生命。

河南省地理学国家重点学科培育经费资助

论水环境与人口健康
Water Environment and Human Health

(瑞士)许靖华 著
秦明周 译

河南大学出版社
·郑州·

图书在版编目(CIP)数据

论水环境与人口健康:英汉对照/(瑞士)许靖华著;秦明周译.—郑州:河南大学出版社,2012.11
ISBN 978-7-5649-1082-2

Ⅰ.①论… Ⅱ.①许…②秦 Ⅲ.①水环境－关系－健康－研究－英、汉 Ⅳ.①X143②R123

中国版本图书馆 CIP 数据核字(2012)第 276886 号

责任编辑	董庆超
责任校对	胡凤杰
封面设计	马 龙

出 版	河南大学出版社		
	地址:郑州市郑东新区商务外环中华大厦 2401 号 邮编:450046		
	电话:0371－86059701(营销部) 网址:www.hupress.com		
排 版	郑州市今日文教印制有限公司		
印 刷	郑州市今日文教印制有限公司		
版 次	2012 年 12 月第 1 版	印 次	2012 年 12 月第 1 次印刷
开 本	890mm×1240mm 1/32	印 张	4.75
字 数	91 千字	插 页	1
印 数	1—1500 册	定 价	26.00 元

(本书如有印装质量问题,请与河南大学出版社营销部联系调换)

目 录

1. Applying Integrated Hydrologic Circuits (IHC) Technology to Resolve Water Resource, Environment and Energy Problems …………… (1)
   应用集成水路方法解决水务、环境及能源问题……
   …………………………………………………… (27)

2. Use of Hydrotransistor and Self-Cleaning Pond for De-nitrification to Produce Purified non-Carcinogenic Water for Drinking and for Groundwater Recharge
   ……………………………………………… (43)
   利用水转换器和自净塘的脱氮作用生产无致癌作用的饮用水和回灌地下水 ………………… (59)

3. A Symbiogenetic Origin of Cancer and the Nitrite Paradox ……………………………… (70)
   亚硝酸盐的特点与癌症共生进化的源起 …… (98)

4. Role of Nitrite in Tumor Growth, Symbiogenetic Evolution of Cancer Cells, and China's Successes in the War against Cancer ……………………… (116)
   亚硝酸盐在肿瘤生长中的作用,兼论癌细胞的共生进化起源与中国抗癌的成就 ………… (137)

1 Applying Integrated Hydrologic Circuits (IHC) Technology to Resolve Water Resource, Environment and Energy Problems

Abstract

The current problems of water shortage and environmental pollution could be resolved through water-recycling and through the balanced use of the groundwater reservoir. The solution of the crisis lies in the finding of a technique to amplify the water flow-rate into and out of porous medium. Hsu, assisted by his volunteer helpers, have devoted the last decade to develop the technology of integrated hydrologic circuits, and they invented hydrotransistors. This is article Part I summarizes the theory and the inventions of hydrotransistors.

R-hydrotransistors serve as groundwater-recharger without loss the use of surface land. RI-hydrotransistors serves a dual purpose of groundwater-recharge and capillary irrigation. F-hydrotransistors can reclaim waste-water for recycling as urban supply or as recharged groundwater. E-hydrotransistors extract groundwater a high enough rate to generate hydroelectricity. The innovations are destined to replace current technology because of their economic superiority. Assisted by legislations, the new technology could pave the way for a Green Revolution.

The forthcoming Part 2 will be case-history studies at Dongguang and at Kunming China to treat polluted and waste waters for recycling to eliminate pollution. Part 3 will report a pilot program to convert desert into crop fields in the Shiyang River Drainage Basin, Northwest China. Part 4 will describe the theory and practice of a patented technology of three-dimensional water-injection for enhanced oil recovery. Part 5 will be a progressive report on groundwater-hydroelectricity.

Introduction

Industrial Revolution came after the French Revolution and the publication of the Malthusian Theory of Population; there was the necessity to resolve the potential crisis of starvation. The industrialization gave us mechanization and power, starting with the invention of steam engines, followed by fuel and electric power. Chemists played a dominant role. Engines were driven by phase and/or chemical changes. Manufacturing industries are purification or combination processes.

The Industrial Revolution created employment opportunities, but the inevitable urbanization destroys traditional values. The IT or the Second Industrial Revolution, came to remedy in part the crisis of alienation and loneliness. Information technology and telecommunication bring us closer again. Physicists played a dominant role. The inventions of transistors and integrated hydrologic circuits have made the constructions of hardware and software possible.

The two industrial revolutions have brought huge benefits to mankind, but the collateral damages of have also been considerable. Coming out of the Second World War, the greed of industrialized nation has led to a third crisis: water and energy shortages and environmental devastations. The time has come for a Third Industrial Revolution, a Green Revolution to make the Planet a better place to live in. Earth scientists will have to play the dominant role. We have to learn from Nature in our search of recipes for clean water, clean air, and renewable energy.

Statement of Problem

Current water and environment crises are a consequence of the misconception of what constitute water resources. In all books, in all practices, water resources are defined as the sum of water volumes in lakes, rivers, and other surface water-bodies. We learned from doomsayers that the World freshwater resources are rapidly vanishing. This is not true, water comes to us in the form of precipitation, and the global rainfall has not been diminishing. In fact, arid regions have received more rainfalls during the ages of global warming. 1 This ultimate water resource has been stored underground as groundwater, and indeed the groundwater reservoir constitutes 95% of the freshwater resource of the world. The natural recharging rate is very small; only 5% to 10% of the rainfall seeps underground to become groundwater, much of the precipitation flows as rivers or canals or stored in lakes or reservoirs on its way to the sea. The small recharge is enough to maintain a balanced water budget. Our groundwater resources are the savings of many million years, and there was no significant deficit spending until the 20th Century. With the building of reservoirs and canals, people became then addicted to the use of surface water. Much water is lost by evaporation and by irrigation in arid region, and by pollution or wasteful human consumption in humid regions. Water shortages have become a norm rather than exception in many parts of the world. We are coming to a stage that there has to be a change in the fundamental policy, if the current crisis is to be remedied. We have to liberate ourselves from the dependence on surface water. On the other hand, the current excessive exploitations of groundwater have caused many problems, such as desertification, land-subsidence, and salinization of groundwater through. Drilling for groundwater is thus restricted or even prohibited in many communities. The solution is not to prohibit, but to regulate.

Water underground is like money in the bank. There are banking laws to regulate the overdraft of bank accounts, and there have to new legislations in order to achieve a balanced exploitation of groundwater. Bank accounts are balanced when incomes are regularly deposited, and the expenses are budgeted not to exceed the incomes. Likewise, the groundwater reservoirs have to be recharged, and use of the water resources has to be regulated by this fundamental economic principle: Those who do not make deposits in the bank are not permitted to withdraw money from the bank, and those who exploit groundwater have to share the expenses of recharging.

The importance of the groundwater recharge has been recognized by numerous scientists, and they have invented various methods of groundwater-recharge technology. As summarized in a study by the British Geological Survey, the current artificial recharge programs are grouped into the following categories:

Spreading methods;
Open wells and shaft;
Drilled wells and boreholes;
Bank infiltration;
Sand storage dams;
Roof-top water.

The category of spreading methods include infiltration or recharge basins, perennial dams, ditch and furrow, flooding, irrigations, stream channel modification, etc. Many large gravel pits are dug, for example, in the City of Beijing or in Orange County California. Flood water is stored in the pits to seep slowly down to recharge the groundwater reservoir. The disadvantages of such spreading methods are manifold. The use of surface land is lost. The clogging of the porous medium tends to hinder the recharging by infiltration. Also there is the substantial loss by evaporation of the water being temporarily stored for recharging. In view of the economics, many urban recharge-pits are filled up, and the land surface is sold as expensive real-estate. The liquidation of recharge-facilities has resulted in ever-more deficit of the groundwater budget. The BGS report has indicated that other recharge techniques are likewise ineffective and costly. Therefore, a greatest task in adopting the use of groundwater resource is to develop an economic and effective recharge technology. Recharging involves the quick entry of surface water underground, and one searches for a solution in finding a technique to amplify the water flow-rate into porous medium.

A major problem of carrying out groundwater recharge program is the lack of surface waters for recharging. The Er-Huang Village program at Beijing, for example, has to be suspended when water from Yunding River was no longer made available. Ultimately, we can trace the urban water shortages to the fact that treated waste-water, in most instances, cannot be recycled for groundwater recharge.

A waste-water has four components

(1) Organic debris, mainly algal remains

(2) Inorganic sediments, mainly sands and mud

(3) Living micro-organisms, mainly bacteria

(4) Dissolved matter, including heavy metals, phosphates, nitrates and nitrites.

One can build plants for aerobic biodegradation to produce intermediate water, zhengshui in Chinese. 3 The organic debris, mainly hydrogen and carbon compounds, are oxidized to carbon dioxide and water. The coarser suspended particles are settled in sedimentation ponds. The living micro-organisms could be disinfected. Only the very fine particles and the dissolved matter cannot be removed. Nevertheless, the treated water has no unpleasant odor and appears scenic, so that the biodegradation method is widely adopted as a standard practice of waste-water treatment.

The method underestimates the deleterious effect of the minor constituents of organic matter. Nitrogen and phosphorous are present at a ratio of about 15:1 in all living organisms. They are released as dissolved constituents, when organic debris are oxidized and decomposed. Added to those are ammonium and phosphates from chemical fertilizers and phosphates from detergents. Since nitrogen salts, especially the carcinogenic nitrites, are harmful, zhengshui is not permitted to be recycled as urban water supply. An unethical practice to use it for irrigation has caused very serious problems of groundwater pollution

An alternative to aerobic biodegradation is to remove organic and inorganic particles by coagulation and settling. Chemicals are introduced for those particles to adhere to the chemical mud before they settle. ? The residence-time of waste-water in sedimentation ponds is, however, much too brief, for reasons of economy. Fine articles tend thus to remain in treated water. The attempt to use the coagulation method for waste-water treatment at Dongguang, China , for example, has failed: the inflow and outflow waters both have strong odor, because much of the partially oxidized organic debris fails to settle out. A critical slogan of drinking water in and stinking water out has caused the City officials to seek a better solution for waste-water treatment.

Still another alternative is to filter out the particulate matter and to absorb the hazardous dissolved matter by io-membranes. Sand filters are used, and water flow under gravity through the coarse filtering material. The disadvantages are two fold, the flow rate is too slow, and the pore size is too large. To increase the flow rate of sand filters, perforated pipes are installed near the base of sand filter, so that filtered water can be pumped out. Still the quality of filtered water leaves much to be desired because of the large pore size the filtering coarse sand.

Micro-membranes and super-membranes have been invented to remove micrometer-size particulate matter, or even some dissolved ions from polluted or treated waters. There are proposals in China , for example, to convert zhengshui into reclaimed water through such micro- or super-filters. The membranes are costly and experiments have encountered numerous problems. Consequently micro-membranes are seldom used for sewage treatment because of the economics.

Still a physical method of filtering is preferred to avoid the production of carcinogenic nitrites, Still. An effective filtering involves the rapid entry and rapid exit of water through very fine granular material. To recycle treated water, one also searches for a solution in a technique to amplify the water flow-rate into and out of porous medium.

Surface waters have been the traditional sources for irrigation. In a study of the water-budget of the Shiyang River drainage system of Northwest China, we found that the constructions of reservoirs and of irrigation canals have been the cause of desertification. Melt water from the mountains and groundwater from karst terrains were stored in reservoirs, transported by canals, and poured into the soil for plant growth. Much of the water is lost through evaporation4. Where surface waters are no longer sufficient, groundwater has to be pumped out for irrigation, causing alarming lowering of the groundwater table and its eventual salinization. The end result is desertification. Fresh or brackish water lakes are converted into salt flats and playas, large tracts of farmland lost, and people deserted their homes. This global catastrophe has alarmed the politicians. The Chinese Government, for example, is planning a 20-billion RMB program to prevent the desertification of the Shiyang River

Drainage Basin. They have recognized the need for a radical policy change. Instead of building more reservoirs and canals for water to be lost to evaporation, or pumping out more water from underground to cause the loss of the precious groundwater resource, there has to be water-saving irrigation technology, and there has to be effective groundwater-recharge technology.

Irrigation is pouring or dripping water into the soil. Plants in natural environments are not irrigated; they grow where water is sucked up by the capillary pressure of the soil from an underground water-bearing stratum, or aquifer. We are calling this process capillary irrigation. A solution of the problem is to construct artificial aquifers under crop fields for capillary irrigation. The solution would be even more attractive, if such aquifers are used for groundwater recharge. Flood water can be filtered through quick entry into, and exit from, a porous medium. The filtered water can then flow, or be pumped, into artificial aquifers where plant-growth is nurtured through capillary irrigation, while the recharged water seeps slowly down to replenish the groundwater reservoir. To maintain a balanced water-budget in arid regions, one needs not only water-saving irrigation, but also a technology to amplify the water flow-rate into and out of porous medium.

Hydroelectric energy is renewable. Water evaporated from the oceans will precipitate in the mountains, and flow back to the sea under gravity. Hydroelectricity has not been considered an acceptable because of the deleterious social and environmental effects of large dams[5,6]. There is, however, no argument against this source of renewable energy if high dams are not necessary. The volume of water in the porous sediments of a silted reservoir is about 40% that of the reservoir, and the volume of water in the thick alluvium of high-mountain valleys above Alpine gorges is also huge. Such groundwater is not extracted for hydroelectricity-generation, because the very slow rate of extraction, and the slow rate of replenishment through recharge. One could utilize groundwater hydroelectricity as a renewable energy if a technology is found **to amplify the water flow-rate into and out of porous medium.**

The foregoing analysis of the current water and environment crises suggests that the panacea to solve all water and environment problems is find a technology to amplify the water flow-rate into and out of porous medium. A first task is to thus find a hydro-amplifier. Recalling that the flow of electricity, when first discovered, was compared to the water-flow, I began 10 years ago to study of the hydrodyanamics of the groundwater flow.

The flow rates of both electric and aqueous flows are proportional to the conductivity, potential gradient, and cross-sectional area, known respectively as the Ohm Law and Darcy Law. With the development of the Integrated Circuits (IC), information technology problems have been solved with the construction of electronic circuit elements, fabricated and associated within a continuous body to perform electronic functions. Perhaps, I thought, we could develop a technology of Integrated Hydrologic Circuits (IHC), consisting of hydrologic circuit elements, fabricated and associated within a continuous body to perform hydrologic functions.

In hindsight, the solution seemed simple indeed: Hsu was looking for something **to amplify the water flow-rate into and out of porous medium.** so that groundwater, instead of surface water, could be the water resources. Since his retirement from the Swiss Federal Institute of Technology, he has devoted himself to develop the new technology7. The most important invention turns out to be the hydrotransistors. Hsu has invented the R-hydrotransistors for groundwater-recharge, F-hydrotransistors for filtering, IR-hydrotransistors for the dual purpose of irrigation and groundwater recharge, E-hydrotransistors for extracting groundwater to generate hydroelectricity. In his endeavours, he received unstinting support from his wife Susan, and financial assistance from his late school-teacher sister Mrs. Leehsia Hsu. Yeh Wenhua, Li Dong, Kong Yunhua, Hu Feng, and Lin Guang are young volunteers for he Cause. We also had the help of many sympathetic colleagues, some of whom are co-authors of this article. Our effort to innovate have, however, met fierce opposition by an alliance of political and scientific establishments.

Physical Basis for the Invention of Hydrotransistors

The invention of transistor is the one invention indispensable to the IT Revolution. Transistor is a resistor or a semiconductor device which can amplify electric signals as they are transferred through it from input to output, from source to drain. In our search for devices which can amplify the rate of water entering into and exit from porous medium, we stumble onto constructions which have amplification and switching on & off functions like transistors. Hsu coined the word hydrotransistor to designate such devices, and consider them the most important component of integrated hydrologic circuits; the other IHC components are hydro-conductors, hydro-resistors, hydro-insulators, hydro-capacitors, hydro-filters, hydropotentials, etc.

A triode transistor has a source, a gate, and a drain, or an emitter, a base, and a collector. The functioning of transistors has been compared to water-flow through a washbasin. There is a source entering the transistor, and the flow rate is regulated by faucet, or the ate? of the transistor. ? Water entering the basin will flow out through the drain. In a steady state flow the rate of water flowing out depends upon the rate flowing in, and the rate is regulated by the opening or closing of the faucet; a larger rate will flow in an out if the faucet is opened wider.

The flow rate of electrons, or the intensity of electric current, is stated by Ohm law

Intensity = Potential difference/ Resistance

Where wires of different cross-sectional area are used

Flow rate / cross sectional area = conductance x potential difference

In using transistors as amplifiers to amplify the flow rate, the key is to regulate the current flow across the gate. Through the accumulation of electrons in a semi-conducting medium that serves as base, the potential difference is increased so that the current flowing from the emitter could be increased up to 10,000 times to flow into the collector.

Readjusting the terms and using the technical language of hydrology, the Ohm Law is translated into Darcy Law of fluid flow through porous medium:

Flow rate = transmissibility x hydraulic gradient x cross-sectional area

In constructing hydrotransistors, we adjust the fluid flow into and/or out of porous medium through constructions which change the transmissibility, hydraulic gradient, and/or cross-sectional area of the fluid flow.

For recharge by hydrotransistors, water flows laterally into a layer of porous and permeable sediments, where it seeps down slowly to replenish the groundwater reservoir. The key change is the variation of the cross-sectional area of the flow. The simplest IR-hydrotransistors are thus a layer of sand or gravels, or an artificial aquifer of a semi-conducting medium, buried somewhere between the surface soil and the fine sediments in the vadose zone. Rainwater, filtered flood-water, or circulating groundwater can flow or be pumped into the artificial aquifer at a high linear rate across the small cross-sectional area of the lateral flow. From the porous medium, water seeps down at a very slow linear rate through the underlying fine sediments, because their very small transmissibility. The volumetric rate could be maintained at a steady rate, if the cross-sectional area of the vertical flow is very large.

The volumetric rate of water flow through a hydrotransistor does not have to be steady, and the law of hydrodynamic continuity does not need to apply. Water could be fed periodically into the artificial aquifer, while it seeps down continuously. This non-steady flow forms the physical basis of constructing IR-hydrotransistor. Water could be injected, or pumped in, quickly into a porous and permeable artificial aquifer, but the downward seep is very slow where the underlying sediments are relatively impermeable. ? The recharged water could stay thus for sometime in the aquifer during its transit journey of recharge. Meanwhile the capillary pressure of the sediments above would draw water up to wet the soil and to nurture the growth the plants. To further reduce the rate of evaporation from the soil, a hydro-insulator is constructed at a shallower depth below the top soil.

The advantages of installing IR-hydrotransistors are thus:

(1) the artificial aquifer serves as a underground storage for groundwater-recharge, without the need of using surface land such as the case of digging ponds or gravel-pits as recharger.

(2) the evaporative loss of the water being recharged from an artificial aquifer buried in a meter or

more under the surface (Figure 1, 50) is much less compared to that of surface evaporation.

(3) the water in the artificial aquifer (in transient) serves to keep the overlying soil or fine sediments (Figure 1, 54) wet to nurture the growth of plants so that there is no need to pour water onto land surface for irrigation.

Where rapid extraction of water from underground is necessary, one could build E-Hydrotransistors which have a function exactly reversed from that of R-hydrotransistors.

Water seeping rapidly underground can flow into an artificial aquifer, from which water is quickly pumped out.? The installation of E-hydrotransistors can thus prevent flooding of homes, of low-lying land on the side of rivers and/or of lake shores. Where high alluvial valleys above river gorges are underlain by thick alluvial sediments, rapid extraction of groundwater by E-hydrotransistors could result in a flow rate rapid enough to turn turbines to generate electricity. The water budget of the groundwater in such valley sediments could be balanced if R-hydrotransistors are installed to compensate for the volume extracted.

Where water is to be filtered, both the rate of flowing into the source and that of flowing out of the collector should be rapid. A rapid lateral rate can be achieved if water is pumped out of perforated pipe, and an equally rapid volumetric rate of seeping through the filter can be achieved if it has a large enough area. Since the flow through permeable medium is governed by transmissibility, conventional filters use coarse sand as filtering material, rather than fine sand or silt which has permeability orders of magnitude smaller. The pore size of such sieves is, however, so large that finest particulate matter cannot be removed.

Our major contribution in pollution-control and waste-water treatment is the invention of F-hydrotransistor. Conventional filters in waste-water treatment consist of water percolating through a layer of coarse sand under gravity. In some plants, perforated pipes connected to pumps are embedded in the sand to enhance the filtering rate. Nevertheless, the quality of filtered water leaves much to be desired. We initiated, therefore, a series of experiments to try the finest filtering material that could still give an economically practical flow-rate.

Our construction has three components, a triode, a filter, and the intervening protective mantle. In order to give the hydrotransistor a maximum potential difference across the filter, we need to create negative pressure i. e., pressure smaller than atmospheric pressure under the filter. The negative pressure can be produced where water flows out of pipes that are connected to pumps. Where the pipes are perforated and are embedded in a gravel matrix, water from the pore space of gravel enters the pipes, and the negative pressure is thus transmitted from the inside of the pipe to the gravel matrix. The pipes constitute thus the rate of the triode, a rapid rate of lateral flow through the pipes is induced by the steep pressure-gradient from one end to another of the relatively short pipes.

The water from the pore space of the over-lying protective mantle seeping into the semi-conducting gravel is the source of the triode. A protective mantle is necessary if a silt is used as the filtering material, because its grain size is smaller than the pore size of the gravel so that silt grains would tend to infiltrate the pores of the gravel. The protective mantle is a layer of sand grading from fine to coarse sand downward. Silt cannot infiltrate the pores of fine sane, fine sand cannot infiltrate the pores of medium sand, medium sand cannot infiltrate the pores of the coarse sand, and coarse sand cannot infiltrate the pores of gravel.

The vertical flow rate through the filter and the protective mantle is obviously related to the transmissibility, which is related to the square of the grain size of the filtering material. Invention is, however, the discovery of the unexpected. We found to our surprise that the flow rate through silt filters is not much slower than that through sand filters. While a sand filter with a grain size 10 times larger than that of a silt filter, we expected a 100-fold difference in flow rate. To our surprise, the difference was on 2- or 3-fold. The finding seemed to contradict the relation given by Darcy Law.

Hsu puzzled over the problem for several months, before it finally dawned on him that there are two laws of fluid flow governing the filtering by hydrotransistors: the Darcy Law and the Law of Hydraulic Continuity. To maintain a steady state flow, the vertical hydraulic gradient across the filter and protective mantle is not linear. The gradient across the silt filter must have been many times greater than that across the protective mantle so that the volumetric flow through the hydrotransistors is a constant. The construction of hydrotransistor is not simply to insert a pump near the base of a sand filter.

An optimum design for economy depends upon a knowledge relating the grain size and thickness of the filtering layer, the grain size and thickness of the sand layers of the protective mantle, the diameter and length of the perforated pipes, and the power of the pumps. There are many possibilities for economy. For example, in our original design to use Hydrotransistor Model 2006A1 to treat 6 million tons of waste-water daily, we expected to build 15,000 such units, each capable of processing 400 tons daily. Such a set-up would cost some 7.5 billion RMB. Recognizing that a larger steady flow-rate could result from the use of a different pump, we increased the pump power of Hydrotransistor Model 2006A2. We found that the processing capacity of each unit was increased to 600 tons daily. Higher flow rate signifies the necessity of building few units, and a reduction from 15,000 to 10,000 units represents of 2.5 billion RMB. We are making other alterations of the design for economy, such as the nature of filtering material, the thickness of filtering layer, and expect still more drastic reductions of the cost. We know that our patented design of hydro-transistors might be copied by unscrupulous makers, but the IP of the economy gained from experiences during the R & D will be the trade secret that gives the Beijing Kenneth Hsu IHC Technology Ltd. a competitive advantage.

Legislation a Priority for Revolutionary Changes

In our analysis of the current problems, we came to the conclusion that current problems have resulted from the mistaken reliance upon surface waters as the water resources. Pollution and evaporative loss have resulted in severe water shortages. The situation is further worsened because the faulty practices of waste-water managemen. The only way out of current water crisis, in my opinion, is a policy that

(1) relies upon the groundwater reservoir as the water resource;

(2) harvests rains and floods for groundwater recharge to achieve balance in water budget;

(3) reclaims treated waste water for recycling as urban supply or to be recharged underground.

The success of implementing those policies depends upon legislations and economic incentives.

City ordinances are now adopted by numerous European countries to regulate that rainwater collected from roof-top has to be separated from sewage. The economic incentive is to reduce by half the charges of sewage-treatment fees. A normal practice to dispose such rainwater is to dig a gravel pit. To replace such pits by RI-Hyotransistors under garden lawns will serve a multiple purpose of disposing meteoric water, of recharging groundwater, and of watering the grass. With new legislations, there should be economic incentives to install large RI transistors under lawns, meadows, parks, golf courses, sport fields, etc. When systematic recharging is carried out in a city, the quantity of water recharged by juristic persons could be monitored and recorded, to be partially refunded in the form of reduced cost of water-consumption as a further economic incentive.

Legislations are required so that treated waste water should not have components that are deleterious to health, so that it can be reclaimed for urban supply or for groundwater recharge. Sewage treatment plants producing water with carcinogenic substances should be shut down, or be remodelled so that no such health hazards could be produced.

With the legislation requiring the installation of IHC systems to balance water-budget, land subsidence of many cities and desertification of arid regions can be prevented. When the filthy canal water of Venice can be filtered by RF-Hudrotransistors and charged into groundwater reservoirs, the City could be uplifted. Where surface irrigation is forbidden by legislation, IR hydrotransistors have to be installed to convert waste desert lands into crop fields or grazing meadows.

Finally, the current practice of building chimneys to dispose of waste gas into the atmosphere is reminiscent of the medieval custom of dumping waste-water onto the streets. There should be legislation to require that factories, if not homes, producing an undue amount of waste gases be responsible for a pipeline system to connect them. There can even be an economic incentive if those releasing their carbon dioxide to acidize polluted water to combat pollution be compensated one way or another for their contribution to alleviate global warming.

References Cited

1. Hsu, K. J. (2000) Klima macht Geschichte. Zurich: Orelll Fueslli Verlag, pp. 334.
2. I. Gale, Neumann, I. Calow, R. and Moench, M. (2002) The Effectiveness of Artificial Rechage of Groundwater: a Review. Keyworth, Notthingham: British Geological Survey Commercial Report CR/02/108N, pp. 51.
3. Tao junji, Yu juntin, Zhen Zhengyuan,? (2005) Chengshi Wushhui Chuli Jishu ji Giongchengshili (Urban Water Treatment Engineering). Beijing: Chemical Engineering Publishers, pp. 332.
4. K. J. Bertels, K. Wang, B. C. Ramaekers, B. (2004) To save a desert oasis through the installation of an experimental integrated hydrologic circuits and rational water management. Rejected proposal submitted to the Framework Program of the EU-China Science and Technology Cooperation, pp. 26.
5. Goldsmith, E. and Hildyard, N. (1984) The Social and Environmental Effects of Large Dams. Wadebridge Bridge, Cornwall: Wadebridge Ecological Centre, pp. 346.
6. World Dam Commission (2000) Dams and Development: A

New Framework for Decision Making. Sterling VA: Earthscan Publications Ltd. , pp. 402.

7. The inventions patented include

KJ 1 USA Patent 6,120,210, Sept. 19, 2000 Use of Porous Medium in an Integrated Hydrologic Circuit for Water Storage and Transport in land Reclamation, Agriculture, and Urban Consumption.

KJ 2 Taiwan Patent 477852, March, 2002.
Integrate Hydrologic Circuit.

KJ 3 Taiwan Patent Appl. 88121914 December, 1999.
Integrate Hydrologic Circuit.

KJ 4 Taiwan Patent 89114962 April 2002 Integrated Hydrologic Circuits in Valley- systems, for Water-supply, Hydroelectricity and Flood Control.

KJ 5 Chinese Patent Application, May, 2005.
Aquitransistors in Integrated Hydrologic Circuits.

KJ 6 Switzerland/PCT Patent Application, PCT/CH 2004/000372, filed June 172004, Aquitransistors in Integrated Hydrologic Circuits.

KJ 7 Switzerland/PCT Patent Application, PCT/CH 03/00482, filed July, 2003 Use of Industrially Produced Carbon Dioxide for the Suppression of Algae and the Rehabilitation of Polluted lakes.

KJ 8 EU//PCT Patent Application, PCT/EP 2006/069183, filed December 1, 2006 Hydrotransistors and Integrated Hydrologic Circuits.

应用集成水路方法解决水务、环境及能源问题

摘 要

当前水资源短缺和环境污染问题可以通过水资源回用和对地下水的合理使用来解决。污水处理的解决主要在于找到一种能够增大水流进出微孔介质的技术。许靖华用了几十年时间,与他的志愿研究成员一起,发明了一种集成水路技术,发明了水转换器。这篇文章阐述水转换器的理论及发明。R型水转换器处理的出水可以用来回灌地下水,RI型水转换器处理的出水可以用来进行地下水回灌和地表灌溉,F型水转换器可以将废水进行回用处理,E型水转换器可以以一个很快的速度将地下水抽出,可以达到水力发电的目的。这些变革性的技术都将能取代目前的技术,因为这些技术在经济上具有目前的技术无法比拟的优越性。如果能得到立法的支持,那么这项崭新的技术将能掀起一场"绿色革命"。

引 言

在法国大革命和马尔萨斯人口论发表后,工业革命

开始。在那个时候,解决人们温饱问题是非常急迫的。随着蒸汽机的发明以及燃料和电气的发明,工业化带给了我们机械化和巨大的能量。在这其中,化学家们起了主要的作用,机车就是通过相变或者是化学能的转变来驱动的,而制造业是一个纯化或者化合的过程。

工业革命产生了很多的就业机会,但是不可阻挡的城市化进程破坏了传统价值。IT或者第二次工业革命在一定程度上缓解了人们日渐疏远的危机,信息技术和通信技术使得人与人之间的距离越来越短,物理学家在这个过程中扮演了最主要的角色。晶体管的发明和集成电路的发明使得硬件和软件的构建和相互链接成为可能。

这两次工业革命给人们带来了很大的好处,但是同时也造成了很大的间接的不良影响。从二战的阴影中走出来的那些工业化国家又面临着第三个危机:水和能源短缺以及环境恶化问题。所以现在进行第三次工业革命的时代已经到来,需要有一场绿色革命来让地球更加适合人类居住。在这个时期,地质学家们会在其中起到最关键的作用。我们需要向自然界学习获取清洁的水源、干净的空气和可再生的能源的方法。

问题阐述

现在的水和环境危机是对水资源构成的误解的后果。在所有的书籍、所有的实践里,水资源都被定义为河湖及其他地表水体中水量的总和。预言家告诉我们,世界上的水资源正在迅速地消失。事实并非如此,水资源以降水的形式来到我们身边，全球的降雨量并没有减少。事实上，在全球变暖期间，干旱地区的降雨量反而会更多。这些水资源最终以地下水的形式储存在地下，实际上地下水的存量占世界洁净水资源的95%。自然回灌占的比率是很小的，只有5%～10%的降雨渗入地下，成为地下水，大部分降水通过河流或渠道流走，或在流向海洋的过程中储存在湖泊或水库里。很小的回灌就足够维持水平衡。我们的地下水资源是数百万年来的积累，在20世纪以前并没有出现明显的超采。通过修建水库和水渠，人们开始增加地表水的使用。在干旱地区，很多水由于蒸发、灌溉而消耗；在潮湿地区，很多水则由于污染和浪费而失去。在世界的很多地方，水资源短缺已经成为一种正常现象而不是意外。要补救现在这种危机，我们必须从根本的政策上进行变更。我们必须从对地表水的依赖中解放出来。另外，现在对地下水的超采又已造成了很多问题，比如沙漠化、地面沉降、地下水盐渍化。因此很多社区都限制甚至禁止了地下水的钻取。解决的办法不是为了禁止，而是为了调控。

地下水就像是银行里的钱，银行有法律去调控账户透支，而为了维持地下水的平衡开采，也必须有新的立法。当收入可以有规律地存储、保持消费开支不超过收入时，银行账户就可以保持平衡。同样地，地下水必须进行回灌，而且水资源的使用也必须按照这个基本的经济原则进行调控：在银行里没有存款的人不允许从银行里取款，同样道理，开采地下水的人必须分担回灌的费用。

很多科学家都意识到了地下水回灌的重要性,并发明了各种地下水回灌技术。据英国地质局报告(BGS)总结,现在的人工回灌项目可分为以下几类:

分散式;

露天井轴;

钻井与地上凿洞;

浅滩渗透;

屋顶水。

分散式回灌包括水池的渗透或回灌、长期的筑坝、沟渠和犁沟、洪泛、灌溉、溪流改造,等等。例如,在北京或加利福尼亚的 Orange County,人们就挖了很多的沙坑。洪水积蓄在坑里可以慢慢地回灌到地下。分散式的缺点很多:占用土地;多孔渗水的媒介一旦被堵塞就会阻碍回灌;临时贮存的水也确实会由于蒸发而损失掉。从经济学角度来看,很多市内的回灌坑都被填满,而土地的表面是宝贵的不动产——土壤。回灌设施的清算造成了更多的地下水赤字。BGS 报告显示,其他回灌技术同样是低效和昂贵的。所以,地下水利用中最大的任务是发展经济有效的回灌技术。回灌涉及地表水的迅速渗透,需要研究的是找到一种提高多孔介质水流率的技术。

实施地下水回灌项目的一个主要问题是缺少回灌用的地表水。例如,北京的西黄村项目,由于永定河来水的不足已经暂停了。因此,我们需要将处理过的污水再生回用于地下水回灌。

污水有四种组分:

（1）有机颗粒物，主要是残留的藻类；

（2）无机颗粒物，主要是沙子和污泥；

（3）微生物，主要是细菌；

（4）溶解物，包括重金属、磷酸盐、硝酸盐和亚硝酸盐。

人们可以通过生物降解的方法获得中水。有机物（主要是碳氢化合物）被氧化成二氧化碳和水。大的悬浮颗粒物在沉淀池中固定，对活着的微生物进行消毒。只有很小的颗粒物和溶解物不能去除。但是，处理过的水没有令人讨厌的气味，看上去也很好，所以生物降解作为标准的污水处理方法被广泛应用。

但是该法低估了溶解物的害处。在所有生物有机体中，氮、磷的比例大约为 15:1。当有机物被氧化降解时，它们变成溶解物释放到水体中。此外，中水还有化肥中的铵、磷酸盐和清洁剂里的磷酸盐。由于氮化合物（特别是有致癌作用的亚硝酸盐）是有害的，中水是不允许回用于城市供水的。不道德地把它用于灌溉已经引起了很严重的地下水污染问题。

好氧生物降解的一种替代方法是利用絮凝和沉淀去除有机和无机的颗粒物。在化学污泥沉淀之前，使用化学药剂使这些颗粒物黏附在污泥上。但是，出于经济的考虑，污水在沉淀池中的停留时间都很短。因此，最细小的颗粒都还留在处理过的污水中。例如，在中国东莞，絮凝的污水处理方法已经被证明是失败的，由于很多部分氧化的有机物没有沉降，进出水都有强烈的气味。对进出污水处理厂的污水的批评已经引起了市政府的注意，他们开始寻找一种更好的污水处理方法。

另一种替代方法是过滤，就是利用生物膜吸附和吸收有害的溶解物。可以使用沙滤，污水由于重力作用流经粗糙的过滤介质。这种方法的缺点有两个方面，即流速太慢和孔径太大。为增加沙滤的流速，可在沙滤底部安装多孔管，这样就可以把过滤后的水抽出来。由于过滤用的是沙子，孔径很大，过滤后的水与要求相差还很大。

人们已经发明了微孔膜和超膜用来去除污水或处理过的废水中的微米级的颗粒物甚至一些溶解态离子。比如，在中国，人们用微孔膜和超膜将中水转化成再生水。这些膜造价高，而且实验中遇到了很多问题。所以出于经济原因，很少将微孔膜用于污水处理。

我们最需要的是一种物理过滤的方法能够避免致癌性的亚硝酸盐的产生。有效过滤包括水的快速进入和迅速退出，通过很细的颗粒材料过滤水中杂质。因此，再生水，有人研究也可以通过提高多孔介质水流率的技术来生产。

地表水是传统的灌溉水源。在一项有关中国西北石洋河流域水量平衡的研究中，我们发现，水库和灌渠的建设成了沙漠化主要的原因。山上融化来的水和喀斯特岩层来的地下水储存在地表水库中，通过渠道输送并用来浇灌作物。很多水由于蒸发损耗了，在地表水不足的地方，地下水就被抽出来用于灌溉，造成严重的地下水超采和地下水盐渍化,最终的结果是沙漠化。淡水湖变成了盐湖和干地,大面积农地丧失,人们被迫离开家乡。这一全球性的灾难向政治家们发出了警报。例如,中国政府正在计划一项200亿人民币的项目来防止石洋河流域的沙漠化。他们已经意识到必须采取激烈的有效的政策变更。必须开发节水灌溉技术和有效的地下水回灌技术,而不是修建更多的水库和渠道使得水资源蒸发损失掉,或者抽取更多的地下水消耗掉宝贵的地下水资源。

灌溉使水流入或滴入土壤中。自然环境中的植物是不用灌溉的,它们生长的地方可以利用土壤的毛细管压力将水分从含水层吸上来。我们将这种过程称为毛细灌溉。解决问题的办法是在耕地下构建人工含水层用于毛细灌溉。如果这种含水层能用于地下水回灌,这种方法将更吸引人。洪水可以在过滤后很快地渗入并保留在多孔介质中。然后,在毛细灌溉的地区,入渗水可以流入或泵入人工含水层。回灌水缓慢地向下渗入,补充地下水存量。因此为维持干旱地区的水分平衡,我们不仅需要节水灌溉,还需要提高多孔介质水流率的技术来回灌地下水。

水电是可更新的能源。从海洋中蒸发的水分形成山区的降水,在重力作用下流回海里。由于大坝对社会和环境的负面作用,水电并不被人们接受。但是,如果不需要建很高的闸坝的话,人们也不会反对这种可再生能源的使用。在淤塞的水库中,淤泥的含水量大约占水库的40%;阿尔卑斯山峡谷上厚厚的沉积物中的含水量也占

40%。由于提取速度慢和回灌补充率小,这些地下水并没有取来发电。如果找到扩大多孔介质水流率的技术,人们可以将地下水水电作为一种可更新能源。

上述对目前水资源和环境危机的分析表明,解决所有水资源和环境问题的万能钥匙是找到提高多孔介质水流率的技术。因此,首要的任务是找到水流率扩大器。电流的运动与水流的运动相似。电流率早已发明,集成水路的目标就是寻找、发展提高水流多孔介质的扩大器。

根据欧姆定律和达西定律,电流和水流的速度与传导系数、电势差和过流截面面积成比例。随着集成电路的发展,电子元件的构建和集成已经解决了信息技术问题。我想,也许我们可以开发一种集成水路技术,它包括各种水路元件,并将其集成来解决水务技术问题。

如此看来,方法看似很简单:许靖华正在寻找一种扩大多孔介质水流率的技术,这样地下水就可以取代地表水而成为水资源。自从从瑞士联邦技术研究所退休后,许靖华就全心全意地投入了新技术的开发。他最重要的发明是水转换器。许靖华已经发明了 R 型水转换器用于地下水回灌,F 型水转换器用于过滤,IR 型水转换器用于灌溉和地下水回灌,E 型水转换器用于提取地下水发电。在孜孜不倦的工作过程中,他得到了妻子苏珊·米兰的大力支持以及他已故姐姐许丽霞老师的经济支持,叶文华、李东、孔云华、胡峰和林广也是他课题的青年志愿者。我们还得到了很多志同道合的同事的帮助。但是,我们的新观点也受到了不少政治和科学组织的强烈反对。

水转换器发明的物理学基础

晶体管的发明是 IT 工业革命中不可缺少的一项发明。晶体管是一种电阻器或者称是一种半导体,这种装置可以在电信号通过它时将电信号放大后输出。在寻找一种可以将从多孔介质中流出的水流流速扩大装置的过程中,我们就曾选用什么样的装置,能够像晶体管那样可以实现放大的功能并像晶体管那样可以方便地关闭和开启。许靖华发明了"水转换器"这样一个词汇来定义这样的装置,并认为水转换器是集成水路技术中最重要的组成部分。集成水路技术的组成还包括水力导体、水力电阻、水力绝缘体、水力电容器、水力过滤器和水势等。

电子流速或者称电子流密度,是由欧姆定律定义的,即:

电子流密度＝电势差/电阻

在这里可能用到不同截面积的不同粗细的电线,所以单位面积的电子流速定义如下:

电子流速/单位截面积＝电导系数×电势差

用晶体管的方法导入电子流时,关键就是要使通过的电子流规范一致,向一个方向运动。通过半导体中电子的不断积累,电势差不断变大,这样从发射器中发射出的电子流密度就能够比入流的电子密度大 10000 倍之多。

同理,将这样的原理应用到水力学中也是一样的。将欧姆定律用达西定律代替,那么通过多孔介质的水流速计算如下：

流速＝渗透系数×水力梯度×过水截面面积

在组建水转换器时，我们可以通过调整水流的"传导系数（渗透率）"、"梯度（势差）"、"过流截面面积"来增加或减小水体"流量"。

用水转换器进行地下水回灌时，水流进入多孔介质层和渗透层，然后缓慢渗透到地下以补充地下水。主要的变化是调整过流截面面积。最简单的 IR 型水转换器是一个砂石滤层，也可以称它为一个人工蓄水层或是半导体介质，这个砂石滤层位于渗流地区的表层土壤和细砂滤层之间。雨水、过滤后的水、循环的地下水能够流入或者用泵打入到人工蓄水层中，这样就能通过较小的过流截面面积达到很高的过流速率。水流通过多孔介质向下流到下层细砂层中，但是速度较小，因为它们的渗透系数很小。如果垂直流的过流截面面积足够大的话，那么就可以获得一个很稳定的体积流率。

通过水转换器的水流的体积流率并不一定要很稳定，并不一定要在水转换器中保持一个稳定的连续水流，水流可以间歇性地进入到人工蓄水层中，然后让它慢慢渗流下去就可以了。这种不稳定的水流是构建 IR 型水转换器的物理基础。水流可以注射，也可以用泵打到多孔介质中和人工蓄水层中，但是接下来的水流流速却会很慢，因为下面一层的滤层是相对不容易渗透的。这样的话，回灌水就需要在那儿停留一段时间。与此同时，上层滤层的毛细管压会使水向上渗透到表层土壤中滋养植物的生长。为降低表层土壤的蒸发作用,需要在表层土壤以下的位置上设置一种水力绝缘体。

布设 IR 型水转换器的好处有如下一些:

人工蓄水层可以作为地下水回灌时的存水区域,不需要在表层地上再用挖塘或者砂坑等作为存水用;

因为回灌是在地面下 1m 或者更深的地方进行的,所以水在回灌到人工蓄水层时的蒸发损失要比表层蒸发损失少很多;

水会在人工蓄水层中停留一段时间,与此同时,上层滤层的毛细管压会使水向上渗透到表层土壤中滋养植物的生长,这样就不需要再对表层进行灌溉了。

在某些领域,有时需要将地下水快速地从地下抽出,这时就需要组建一个 E 型水转换器。E 型水转换器实际上在功能上是与 R 型水转换器相反的。

快速下流到地下的水能够流入人工蓄水层中,然后水可以被快速地用泵抽出。E 型水转换器的布设可以防止在湖边、河岸等低洼地带的房子受洪水泛滥的影响。通过 E 型水转换器的快速抽取,可以提高水流流速并可以推动涡轮发电。利用 R 型水转换器可以补偿抽取出的水量,这样就能够实现地下水利用和回灌的平衡。

在水通过过滤层的时候,流入的速度和流出的速度都应该很快。在用泵将水抽出多孔管时,可以得到很大的横向流速度。同时,如果过流截面面积足够大的话,也同样可以得到相同大小的经过过滤层向下渗流的水流的体积流速。由于通过渗透介质的水流受渗透系数控制,常规的过滤装置都会选用粗糙的的粒径较大的砂子来作为过滤介质,而不会选用渗透系数要小很多的细砂或者泥沙来作为过滤介质。但是那些粗糙的过滤介质的孔径很大,很多细小的颗粒物质都不能被过滤去除。

我们在污染控制和废水处理上的主要工作是发明了F型水转换器。常规的过滤方法主要是使水以重力自流的形式通过一个粗糙过滤介质层。在很多污水处理厂中,多孔管与泵相连埋在滤料底下以加大过滤速率。但是,这些过滤装置处理后的出水的水质仍然有待提高。所以我们经过一系列试验研究,尝试并找到了最佳的过滤材料,既能保证出水水质,又能保证较大的处理流速,这样使得该方法在经济上具有可行性。

我们的装置由三部分组成,即收集部分、过滤器和间于其中的保护层。为了能使水转换器达到最大的水势差,需要在滤层底部产生一个负压(即小于大气压),而负压能通过泵不断地将水抽出到管外。埋在过滤介质底部中间位置的出水管是带有很多小孔的,通过过滤介质之间的空隙流下来的水再通过管子上的小孔流入到出水管道中,管道中的水被泵抽走,这样就在过滤介质和出水管之间形成负压。因此,不同位置之间的负压使得管道中的水流有快速的水流速率。

收集系统的水流来源是从上层保护层微孔处进入半导体式砂滤层的水。保护层对于一个以泥沙作为过滤介质的过滤系统是必需的,因为泥沙的孔径比砂石过滤层介质的孔径要小,所以泥沙就容易进入到砂石过滤层中。

而保护层就是一个由从细到粗不同粒径砂子组成的滤料层，防止泥沙进入到中间的砂石滤料层中，中间滤料层的滤料也就不会再进入到粗糙砂石组成的滤料层中，粗糙砂石滤料层的砂石也不会进入到砂砾层的孔隙中。

通过过滤层的垂直水流速率和保护层与渗透系数有很大的关系。这个发明的产生当时实际是有些在我们意料之外的。我们惊讶地发现，通过泥沙的过滤速度不比砂石作为过滤介质时的过滤速度小多少。砂石的孔径大小是泥沙的孔径大小的10倍，这样按照计算水流速度应该是100倍的关系。但是，结果却只有2到3倍的样子。这个关系似乎和达西定律有一些冲突。

经过几个月的思考和求证，许靖华解决了这个问题。许发现有两个水力流速控制水转换器的过滤过程：达西定律和水力持续流定律。为了取得一个稳定的水力流速，通过过滤层和保护层的垂直水力梯度是非线型的。只有当通过泥沙过滤层时的梯度是通过保护层时的梯度的好多倍时，才能使水转换器的体积流率是一个常数。所以，组建一个水转换器不只是将一个泵放到砂石过滤层底部那么简单。

在经济上可行的最好设计方式决定于过滤介质的孔径和过滤层的厚度的关系，过滤介质的孔径和保护层的厚度的关系、多孔管的直径和长度的关系以及泵的功率大小。所以可以有很多种根据不用经济需要制订的方案。比如，在我们2006年设计的用来处理6000000吨/天的流量的2006A1型水转换器模型中，我们就需要建造

15000个单元处理器,每一个处理器可以处理流量600吨/天。这样的一整套装置需要约75亿人民币。由于选用功率较大的泵能够产生更大的稳定的水流流速,我们在2006A2型水转换器中换用功率更大的泵。我们发现,这时每一个单元处理器的处理能力提高到了每天处理600吨。处理流速越高,就意味着需要的单元处理器的数量就越少。这样,只用10000个单元处理器就可以达到每天处理6000000吨的目标,这样在经济上只用25亿人民币就可以了。我们还可以从经济角度出发,对水转换器进行其他的一些设计上的变化和改动,比如改变过滤介质、过滤层的厚度等,以达到降低处理费用的目的。我们知道,我们的水转换器的专利设计有可能被一些不道德的仿冒者仿制,但是一直以来的研究和实践经验使我们北京许靖华集成水路技术发展有限公司在以后的竞争中一直处于有利的位置。

对本革命性的新技术做立法上的支持

在对目前存在的问题的分析中,我们可以发现,目前的问题主要是错误依靠地表水作为可以利用的水资源。污染和蒸发损失造成了严重的水资源短缺问题,并且错误的污水管理方式使这种问题更加严重。在我看来,解决这些问题的唯一出路在于建立如下的一种政策:

（1）这种政策决定主要依靠地下水作为可以利用的水资源;

（2）这种政策将雨季和洪水时的水作为地下水回灌

来实现地下水利用和回灌的平衡；

（3）这种政策要求只有经过有效处理的水才能够作为城市供水回用或者被回灌地下水。

这些政策有效成功地实施还需要立法的支持和经济上的激励。

很多欧洲国家的城市法律规定，从屋顶收集的雨水要和污水分开，并且他们用少收一半的污水处理费用这样的方式作为将雨水和污水分开的规定执行的经济激励。一般处理这样的雨水的方式是挖一个沙坑来放置这些雨水。在草坪底下用 RI 型水转换器代替这样的沙坑，可以起到处理降雨、回灌地下水和浇灌园林植物多种作用。新的立法应该用某种手段的经济刺激来鼓励人们在草坪、牧场、公园、高尔夫球场、运动场等场所下面布设 IR 型水转换器。当一个城市的地下水回灌体系建立起来，回灌到地下的水量就可以被追踪和记录下来，这样能部分地减少水消耗费用，也能够实现以后更长远的经济激励。

需要立法规定处理后的水以及城市供水或地下回灌水不能含有某些对人体有害的物质。在污水处理过程中产生致癌物质的污水处理厂应该被关闭或者进行重新整改，直到不产生任何有害物质为止。

如果立法使用集成水路系统，那么就可以实现水资源利用和补充的平衡，解决很多城市面临的地面下沉的问题以及某些地区的土壤荒漠化问题。当威尼斯的不干净的运河水能够用 R 和 F 型水转换器过滤处理并回灌到地下水中,这座城市可以将自己再提升一个档次。在某些不可以进行表面灌溉的地区,IR型水转换器可以将那些荒漠化的地区变成绿洲或者可以放牧的草场。

最后,目前这种通过管道将工厂产生的废气排入大气中的情况,就像在中世纪随便将污水倒在大街上那样的令人讨厌。所以我们应该立法去要求那些产生过多数量的废气的工厂负责将它们的废气进行收集。当然,也可以给那些用收集的CO_2来酸化污水的工厂一些经济上的补偿,以表彰它们对减轻污染和全球气候变暖的贡献。

2 Use of Hydrotransistor and Self-Cleaning Pond for De-nitrification to Produce Purified non-Carcinogenic Water for Drinking and for Groundwater Recharge

Abstract

Water recycling should be the key to resolve the current water crisis, but the nitrite in water is the stumbling block. Dissolved nitrite salts cannot be removed on a large scale by current technology. On the contrary, waste-water treatment plants, relying on biodegradation process to remove particulate organic matter, produce large quantity of the pollutant so that the treated water is nowhere allowed to be recycled. Current practices of illegally recycling the treated water for irrigation and other purposes have led to catastrophic pollutions.

We have invented filter-hydrotransistor to remove particulate matter in polluted or waste water, so that the biodegradation is not necessary in sewage-treatment. We have further invented a method to remove the dissolved nitrite through the introduction of carbon-dioxide to self-cleansing pond in which diatoms can grow. They utilize nitrite in photosynthesis and they aquatic animals. The fecal pellets of fish sink to water bottom so that nitrite is removed from hydrosphere in a sedimentation process, while an equivalent amount is released by erosion for a steady-state nitrogen cycling on Earth.

The inventions to produce purified, non-carcinogenic water economically on a large scale serve (a) to eliminate pollution, (b) to produce water for recycling as water-supply or as groundwater recharge, and (c) to produce inexpensive and healthy drinking water for everybody.

Health Hazard of Nitrate and Nitrite

Nitrate itself is generally not harmful to human. However, digestive processes in human body can transform nitrate into nitrite. Nitrite is a very strong reducing agent. Absorbed in the blood stream, nitrite is oxidized through its reaction with the oxygen in blood, and causes the blood vessels to lose their functions to transport oxygen.

Baby blue syndrome (methemoglobinemia) is caused by exposure to very elevated levels of nitrite in infants, but the disease is rarely diagnosed in the US. A far more serious hazard is the transformation of nitrite into nitrosamines, and those are proven carcinogenic in laboratory animal studies. It has been pointed out, however, that nitrite could react with the gastric acids in human stomach, so that nitrite could be eliminated before absorption take place. 2 The alarmists found, however, by experimentation gastric absorption of nitrite in animals. Furthermore, nitrite is present in human excrements, signifying that nitrite is absorbed into the human blood stream before its completely oxidation in stomach. Those clinical arguments give credence to the conclusion of numerous statistical studies that correlate cancer mortality to nitrite in drinking water.

Since the issue of nitrite hazard is still somewhat controversial, there is no consensus on the MCL (maximum contamination level) of nitrite in drinking water. The goal (MCLG) is, of course, nil for all potentially carcinogenic substances in drinking water, but Zero-Nitrite tolerance is not practicable, because there has been no treatment-technology to mass-produce de-nitrified water. There is no international standard on the permissible or PMCL; it is different in different countries. The very high PMCL of 1 mg/L NO_2^- in the United States has been set on the basis of early clinical studies, going back to 1950s. The UK authorities are equally relaxed about the potential risk: The PMCL for England and Wales was in fact raised in 1999, from 0.1 to 0.5 mg/L NO_2^- (or 0.033 to 0.165 mg/L of N in NO_2^-). The MCL values of nitrite in drinking water of USA are high; those values do not exceed the PMCL of the United States, but many exceed that of Europe.

The MCL values of drinking water in China are high. The nitrite content of Pearl River (source of Canton's water-supply) is, for example, not only high but steadily increasing, from 0.41 mg/L NO_2^- in 1991, to 0.73 mg/L NO_2^- in 1998, a 50% increase in 7 years. The cancer mortality increased from 1240 to 1540 per million population, an increase of 25%. In other words, some three thousand more persons died of cancer in the City of Canton during those 7 years. How many of those died because their drinking water is carcinogenic?

Unable to solve the problem, the nitrite content of the Pearl River continues to rise to about 1 mg/L NO_2^- in 2006.5 How many more died when the nitrite was increased by another 50%? It was discovered that the tape water had even more nitrite, with values up to 1.4 mg/L NO_2^-, exceeding thus even the very tolerant PMCL of the United States. The addition was related to the liquid-chlorine method of sterilization, which produced even more nitrite.

Foreign correspondents cited the increase of the cancer mortality, and blamed the Chinese Government: the Chinese are "paying the price for decades of environmental neglect. The accusation missed the mark. Experts in China are aware of the nitrite hazard, but the government is facing a dilemma that is faced by all governments of the World. The United States Government set a very tolerant standard, while refusing to accept the statistical correlation of cancer mortality and nitrite concentration. The Chinese Government takes a different approach. There would be no urban water supply, if an acceptable standard is set, as long as there is no technology to de-nitrify drinking water on a large scale. Under the condition, people can either take the risk or they drink "bottled purified drinking water." For that, there is a very stringent standard. The PMCL in 1995 (GB 8537-1995) was 0.005 mg/L N in NO_2^- (or 0.0165 mg/L NO_2^-), and the PMCL in 1998 (GB 17324-1998) was reduced to 0.002 mg/L N in NO_2^- (or 0.0066 mg/L NO_2^-). Those values approach that of Zero GMCL, and are far less than the PMCL of Europe or that of the United States. "Purified drinking water" in bottles or in barrels of 19 liters each are sold in Chinese supermarkets. Osmosis is the method of de-nitrifying such "purified water".

Source of Nitrite from Pollution

Recent nitrification of natural waters is anthropogenic. A major culprits is the disruption of the Nature's recycling. There is no nutrient-enrichment in pristine environments.

Natural waters in mountains are, as a rule, slightly acidic. Erosion and stream transport bring nutrients N and P to mountain lakes, where the dominant plankton are the diatoms. The nutrients are taken up by the plankton, which feed aquatic animals. Fish excrements in the form of fecal pellets sink to water bottom, so that the lakes can remain clear and transparent and low in N and P.

Lakes in lowland, especially in terrains underlain by limestone, tend to be alkaline. Lake water saturated with dissolved calcium bicarbonate ions keeps the lake alkaline; the lake water has an pH of about 8.3 where it is equilibrated with atmospheric carbon dioxide. The condition was, however, different in the spring, when bottom water is brought up by lake circulation. Under such normal circumstance, diatoms grow in the nutrient-rich, slightly acidic water. The diatoms blooms consume much of N and P, so that when green algae should start ? to grow in early summer, the lake becomes impoverished in nutrients. The population of the green algae are thus limited.

The conditions are drastically changed when sewage

or treated waste-water enters a lake; the water becomes more alkaline. The increased alkalinity inhibits the growth of diatoms. More green algae would grow, to make the water even more alkaline, further favoring the blooming of the green algae. Being slightly toxic, green algae are not consumed by the fish, so that the algal population could continue to flourish until the plankton die in the winter. Their dead remains sink then slowly to lake bottom, extracting oxygen from the water column while organic debris decay. The biogenic nitrogen is oxidized in part into nitrate and nitrites and dissolved in water to make the lake even more nitrified when streams bring in more nitrogen salts. The chain reaction is thus alkalinity, growth of green algae, more alkalinity, more green algae, death and decay, oxidation of dead remains, eutrophication, and release of nutrients, such as the harmful nitrite. After "devil's cycle" continues for decades, the lake water with its nitrite hazard can no longer be suitable water-supply. Deprived of a natural process to cleanse the water every spring when diatoms grow, lakes become increasingly alkaline sewage pits for the blooms of polluting algae.

    The one significant environmental difference between the diatoms and the green algae is that fish feed only on diatoms, not on green algae. The nutrients consumed by diatoms are to be embedded in sediments, and the nutrients consumed by the green algae go back to the water column.

Source of Nitrite from Sewage Treatment

Sewage-treatment plants are another major source for the nitrite in drinking water.

The four components of pollution are the inorganic and organic detritus, the living micro-organisms, and the dissolved components. Sewage-treatment plants use screens, sand filters, and sedimentation ponds to remove particulate matter. The rate is very slow, and such physical treatment are not effective. The current practice emphasizes biodegradation to remove organic debris. Organic matter is decomposed by bacteria and is converted into water and carbon dioxide, while the minor constituents N and P are released. Biodegradation under reduced condition produces harmless nitrogen, but the process is too slow and costly so that the more efficient biodegradation by oxidation is commonly employed. Specialists in sewage-treatment seem to be oblivious of the nitrite hazard, and they are not concerned with the consequence of mass production of nitrite in treated water.

The government prohibits the recycling of treated

water zhongshui, but few seem to understand that the regulation is necessary to minimize the nitrite hazard. Nitrite PMCL is not on the list for treated water, so that few Chinese treatment plants bother to make nitrite analysis. Our few analyses show no significant reduction of nitrite after treatment.

In their ignorance of nitrite hazard, water-treatment managements are aggressively promoting the re-use of zhongshui for irrigation. They do not seem to understand that nitrite salts remain in soil after the evaporation of the irrigating water. Studies have shown that the nitrite content in the soil of fields irrigated by recycled wastewater is twice as high as that of fields irrigated by freshwater.6 The nitrite in soil is then dissolved by the next rain to pollute streams and groundwater.

Filter Hydrotransistor and Selfcleansing Pond

To resolve the current nitrification crisis, we have

(1) invented a filter with a very small sieve size and very fast flow-rate to effectively remove particulate matter so that sewage-treatment by biodegradation is not necessary;

(2) invented a process to remove dissolved nitrite salts from filtered water;

(3) invented a process to change the natural environment so that the filtered water would not again be polluted by green algae.

Hsu invented filter-hydrotransistor (FH) to remove particulate matter, inorganic, organic, or living from polluted water. For effective filtering, the top layering of filtering grains consists of medium or fine silt. For fast flow-rate, a hydrotransistor consisting of perforated pipes embedded in a gravel matrix is constructed at the bottom. The filter and the hydrotransistor are separated by a protective mantle of graded sand which prevents the penetration by fine filtering- grains into the large pores of the gravel.

The perforated pipes are connected to pumps, which produce a negative hydrodynamic potential for the water in the gravel. The potential gradient is, however, very gentle from the hydrotransistor to the bottom of the filter. but it is very steep between the bottom and the top of the filtering layer. The hydrodynamic potential gradient is self-adjusting to permit a hydrologic continuity of the flow.

With such a set-up water flows very rapidly through micrometer pores of the 10-micron silt at a rate of about $10 \text{ m}^3/\text{m}^2$. Where the quality of treated water is of no great concern, very-fine sand filters are used to produce scenic water with filtering rate of more than 25 m3/m2. The treated water is clear, devoid of suspended matter, and is in almost all aspect a Grade I water suitable for recharging, except for the dissolved N and P salts (see Table).

Concentration of Nutrients in mg/Lin Treat Water Samples

	A	B	C
Total N	18.89	2.08	1.24
NH_3 N	1,60	0.12	0.11
Nitrate-N	9.85	0.1	0.0032
Nitrite-N	0.15	0.06	0.0004
Total P	0.23	0.068	0.346

Sample A is the analysis of a treated water sample fro a Beijing water-treatment plant.

Sample B is the analysis of a sample of water flowing out of a hydrotransistor.

Sample C is the analysis of a sample of water after two months in a self-cleansing pond.

Filtering is a physical process. Some dissolved ions attached to suspended grains are removed, but chemical or biochemical process has to be invoked to purify chemically. To remove nitrites, three methods are technically feasible: by distillation, by ion-exchange, or by osmosis. Of the three, osmotic membranes are used to produce "bottled purified drinking water" now sold at the Beijing supermarkets, but the process is not economically feasible to de-nitrify water-supply on a large scale.

In our discussion of the source of nitrites, we have indicated that plankton in natural waters can effectively remove nutrients from natural waters. Algae grow by taking up ammonium and phosphate to photosynthesize to photosynthesize carbon dioxide and water.

$$106CO_2 + 236H_2O + 16NH_4^+ + HPO_4^{2-}$$
$$= C_{106}H_{181}O_{45}P + 118O_2 + 171H_2O + 14H^+$$

The ammonium consumed by photosynthesis is compensated by de-nitrification:

$$NO_2^- + 5H^+ + 3Cl^- + H_2O = NH_4^+ + 3HClO^-$$
(acidic environment)

$$NO_2^- + 3Cl^- + 3H_2O = NH_4^+ + 2OH^- + 3ClO^-$$
(alkaline environment)

Therefore, all plankton, diatoms in acidic and green algae in alkaline environment, can remove nitrite from natural waters. Still green algae are considered the very culprit of pollution. Why?

As I indicated, the critical cause of pollution is not the growth, but the death of green algae. Being slightly toxic to fish, they are not a fish-feed. After they die a natural death in winter, their remains sink to water bottom, while the organic nitrogen is converted to ammonia, nitrate and nitrite. Whereas the partially decomposed matter gives odor, the nitrates and nitrites present the health hazard.

Fish do feed on diatoms. We have succeeded to culture diatoms in self-cleansing pond to sustain a population of shrimps and fish. After about a month, nitrates and nitrite are both removed from the filtered water (Table). The nitrite concentration is less than 0.004 mg/L N in NO_2^-: the filtered water has become a "de-nitrified drinking water". The production cost is very little, as the de-nitrification has been accomplished in the self-cleansing pond.

To achieve the aim that the environment is not again polluted by green algae, we have to recall that photosynthesis of diatoms takes up CO_2, N and P. While nitrite is removed, the water in self-cleansing pond is rendered more alkaline. Continued dissolution of CO_2 is thus necessary to maintain the slightly acidic environment to culture the growth of diatoms.

To inhibit the growth of green algae, the pH of the water should be less than below 7. All green algae are exterminated when pH of the water falls below 5.8, but the killing of algae would leave the dead remains, which have to be filtered out. Otherwise, the decomposition of release nitrite again into water.

The sources of the carbon dioxide can be carbon dioxide in tanks, carbon produced by acidization of carbonate minerals, or carbon dioxide collected by a pipeline system from factories burning fossil fuels. The use of carbon dioxide emission to combat pollution of natural waters serves a dual function of alleviating the global warming by greenhouse gases.

Conclusion

Our technique of combining filtering by hydrotransistor and de-nitrification in self-cleansing pond is a solution to reclaim waste water for recycling. The treated water could be directly recycled for human consumption, or it could be recharged underground. With the adoption of such a policy for waste-water treatment and groundwater recharge, there should be no environmental pollution nor water shortage anywhere; waste land and desert can be converted into crop acres and cattle pastures.

References Cited

1. Center for Health Effects of Environmental Contamination, University of Iowa. Report. <http://www.cheec.uiowa.edu/nitrate/faq/html>
2. UK Drinking Water Inspectorate Information letter 15/99- 10 December 1999. <http://www.dwi.lgov.uk/regs/inforlett/1999/info1599.htm>
3. see Mirviss, S. S. 1995. Role of N-nitroso compounds and N-

nitrosation in etiology of gastric, esopharyngeal and bladder cancer. Cancer Lett. , 1995; Sander, J. , 2001Assocation between gastric cancer mortality and nitrate content of drinking water. Eur. J. Epidemiol. , 2001; Gullis, A. 2002. An ecologic study of nitrate in municipal drinking water and cancer incidence in Trnava district. Slovakia, Environ. Res. ; Coss, A. , Canton, P. , Reif, J. S. , Lynch, C. F. , and M. H. Ward 2004, Pancreatic cancer and drinking water and dietary sources of nitrate and nitrite. Am. J. Epidemiology, 159: 693-701, etc.

4. Iowa- Nitrate Fact Sheet. Water report 2003. Pocahontas City, Iowa. < http: //www. pocahontasiowa. com/water-report-2003. htm>

5. Shi, D. , 2006. Chemistry and I, Selected essays of Canton University Science Competition.

6. China's 'cancer villages' pay price : The Chinese villages paying the price for decades of environmental neglect. < http;// news. bbc. co. uk/go/em/fr/-/1/hi/world/asia-pacific/ 627 1103. stm >

7. Masters, Y, Laughlin, R. J. , Stevens, R. J. , and Shaviv, A. 2004. Nitrite formation and nitrous oxide emissions as affected by reclaimed effluent application. J. Environ. Qual. 33: 852-860; Costa, 2002. Nitrate contamination of aquifer and accumulation in the unsaturated zone. Agricultural Water Management.

8. The relevant patents are (a) Switzerland/PCT Patent Application, PCT/CH 2003/00482, July 2003. Use of Industrially Produced Carbon dioxide foir Supression oa Algae and Reha-

bilitation of Polluted Lakes; (b) Switzerland/PCT Patent Application: PCT/CH 2004/000372, June 2004. Aquitransistors in Integrated Hydrologic Circuit; (c) Switzerland/PCT Patent Application: PCT/CH 2006/000002, January 2006. Process for Combating Water Polluted by Algae.

利用水转换器和自净塘的脱氮作用生产无致癌作用的饮用水和回灌地下水

摘　要

解决目前的水资源危机的一个重要方法是进行回收利用,但是水中的氮的存在是进行有效回收利用水资源的一个瓶颈。水中绝大多数的溶解性亚硝酸盐在现有技术条件下不能被有效去除。另外,主要用生物降解的方法去除水中有机污染物的污水处理厂反而产生了很多新的污染物质,使得目前经过这些污水处理厂处理的污水不能实现回收利用。目前这些经过污水处理厂所谓"处理"的出水被用于灌溉等其他用途,这种错误的、非法的回收利用已经造成了很严重的污染问题。

我们发明了一种过滤型水转换器来去除污染水体中的颗粒污染物质。这种方法的应用可以取代目前使用的生物降解的污水处理方法。同时,我们进一步发明了一种利用硅藻去除溶解性亚硝酸盐的方法:我们将 CO_2 引入到自净塘促使塘中硅藻生长,而硅藻通过光合作用进一步利用了水中的亚硝酸盐。水中的鱼类等水生动物以这些硅藻为食,它们的排泄物会沉到水底,这样就能够将水中的亚硝酸盐转移到水体沉积物中,进一步以稳定的氮气的形式回到地球大气圈中,实现水中亚硝酸盐的去除。

这种生产无致癌作用的净化水的方法是非常经济可行的。首先,这种方法可以有效地减少和去除污染,其次,这种方法产生的净化水能回用或者用于地下水回灌,再次,该法可以给人们提供便宜、健康的饮用水。

硝酸盐和亚硝酸盐对人体健康的危害

硝酸盐本身基本上对人体无害,但是硝酸盐经过人体的消化作用可以转变成亚硝酸盐,而亚硝酸盐是一种很强的还原剂,一旦亚硝酸盐进入人体血液后,亚硝酸盐可以和血液中的氧气发生反应,造成血液失去输送氧气的功能。蓝婴症(高铁血红蛋白症)就是由于婴儿接触了高浓度的亚硝酸盐造成的,目前这种病在美国确诊的还很少。更严重的危害是亚硝酸盐转变成亚硝胺类物质,而亚硝胺类物质经过动物试验证明是具有强烈致癌作用的。通过研究发现,亚硝酸盐会和人体胃中的胃酸发生反应,所以亚硝酸盐能在被人体吸收前消除。但也有研究者通过动物试验表明,也存在胃对亚硝酸盐的吸收。同时,在人体的排泄物中也发现了亚硝酸盐,这个结果表明,在人体将硝酸盐完全氧化去除之前,亚硝酸盐还是有一部分进入了人体的血液循环中。大量统计数据的结果表明,癌症死亡率和饮用水中的亚硝酸盐是有相关性的,而很多临床实例更使人们相信这一点。

由于亚硝酸盐的危害问题还存在一些争议,目前饮用水中的最大亚硝酸盐污染浓度还没有一个为大家广泛认可的标准。当然,这个最大亚硝酸盐污染浓度最好是零,使饮用水中的可能致癌的物质均不存在,但是就目前的技术水平而言,零亚硝酸盐是不现实的。目前对饮用水中亚硝酸盐的限制浓度或最大亚硝酸盐允许浓度还没有一个国际标准,美国基于一些早期的临床实例研究将最大亚硝酸盐允许浓度定在 1mg/L NO_2^-。英国政府1999年将英国和威尔士的最大亚硝酸盐允许浓度从0.1 mg/L(0.033 mg/L,以 NO_2^- 的 N 计)提高到 0.5 mg/L(0.165 mg/L,以 NO_2^- 的 N 计)。美国的饮用水最大亚硝酸盐污染浓度非常高,这些值都没有超过美国的最大亚硝酸盐允许浓度,但是有很多已经超过了欧洲的标准。

中国的饮用水亚硝酸盐最大污染浓度也很高。以珠江(中国广州的饮用水水源)为例,1991年亚硝酸盐最大污染浓度为 0.41 mg/L NO_2^-,1998年为 0.73 mg/L NO_2^-,七年中增加了50%之多,亚硝酸盐污染不仅严重,而且有不断恶化的趋势。同时,癌症死亡率也从1240人/百万人口增加到1540人/百万人口,增加了24%。在短短的七年中,广州的癌症死亡人数增加了3000多人。那么,其中有多少是因为饮用了含有致癌物质的水而引发癌症的呢?

由于不能有效地解决这个问题,珠江水中的亚硝酸盐在2006年增加到了 1 mg/L NO_2^- 这样的浓度。如果亚硝酸盐浓度再增加50%,那么会再有多少人死于癌症呢?研究表明,管道水的亚硝酸盐的浓度更高,能达到

1.4mg/L NO_2^-，已经超过了美国的最大亚硝酸盐允许浓度。管道水中的这部分亚硝酸盐的增加应该是与氯化消毒管道水的方法有关,氯化消毒实际上产生了更多的亚硝酸盐。

  国外的记者引用中国癌症死亡率增加的数据,指责中国政府"对几十年来对环境污染的忽视付出代价"。中国的专家意识到了亚硝酸盐的严重危害问题,中国政府也面临着世界上任何一个国家都会遇见的困境。美国政府对亚硝酸盐制定了一个比较宽松的标准,同时美国政府也不接受癌症死亡率与亚硝酸盐浓度有相关性这样的说法。而中国政府采取了另外一种办法。由于目前还没有去除水中的绝大多数氮的方法,如果中国政府制定了一个可以接受的标准,那么中国将无可以利用的城市水资源。在这种情况下,中国人只能承受亚硝酸盐危害的风险或者每天饮用"瓶装饮用纯净水"。鉴于这样的原因,目前中国的标准还是比较严格的。1995 年在中国国家标准 GB 8537-1995 规定的最大亚硝酸盐允许浓度是 0.005 mg/L(以 NO_2^- 中的 N 计)或者 0.0165 mg/L(以 NO_2^- 计)。而 1998 年的中国国家标准 GB 17324-1998 则将最大亚硝酸盐允许浓度进一步减少到 0.002 mg/L(以 NO_2^- 中的 N 计)或者 0.0066mg/L(以 NO_2^- 计)。这些指标有些接近于零最大亚硝酸盐污染浓度,而且比欧洲或者美国的最大亚硝酸盐允许浓度要低很多。瓶装的或者每桶 19L 的"饮用纯净水"在中国各大超市有售,而生产这些"饮用纯净水"的脱氮方法是反渗透的方法。

来自污染物的亚硝酸盐

目前,对水体的硝化脱氮作用是人为地实现氮的去除。这种方法实际上是对自然界的自然循环的一种破坏。而在原始的、未受人为影响的环境中是不会出现氮的富集这种现象的。

自然水体一般都是微酸性的。腐蚀作用和风化作用将 N 和 P 带到山脉的湖中,而在这些湖中的主要的浮游生物是硅藻。硅藻利用了这些 N、P 等营养物质,而鱼类等水生动物以这些硅藻为食。鱼类的排泄物以粪便的形式沉淀到水底,这样就使得水体能够保持清洁透明,并将 N 和 P 维持在一个相对较低的浓度上。

处于较低地势的湖,尤其是在低于石灰岩的地势的湖,通常是碱性的。这些湖的湖水饱含溶解性的钙离子和碳酸氢根离子,使得湖水呈碱性。与大气中的 CO_2 达到气液平衡后,这些湖水的 pH 为 8.3 左右。但是,在春季的时候,情况就有些不一样了。春季由于湖水的循环搅动,底部的湖水被搅动上翻。在通常的情况下,硅藻在富含氮的、微酸性的水中生长。大量的硅藻生长能消耗很多的 N 和 P 等营养物质,所以在绿藻开始生长的初夏季节,湖水已经呈现贫营养状态,这样,也能够限制绿藻的污染问题。

当污水或者处理后的污水进入一个水体后,该水体的条件将大为变化,水体变成碱性的。而增加的这些碱度抑制了硅藻的生长,同时,绿藻在这样的水体中不断生长,使得水体进一步碱化,进一步地促进了绿藻的生长和水华的形成。因为绿藻是有一点毒性的,所以鱼类也不会食用这些绿藻,使得这些绿藻的数量急剧增加,直至冬季死亡。这些浮游藻类的残骸会慢慢地下沉到水体的底部,当这些藻类的有机成分开始分解的时候就会消耗水中的氧气。其中的氮一部分氧化成硝酸盐和亚硝酸盐,慢慢地使得水中的硝酸盐和亚硝酸盐越来越多。这个过程的主要路线为:碱性水体——绿藻生长——碱度增加——绿藻剧增——绿藻死亡——绿藻残骸分解耗氧——富营养化形成。富营养化时藻类的营养物质尤其是有害的亚硝酸盐被释放到水体中。经过几十年这样的"恶性循环"后,水体中将会有很多的有害的亚硝酸盐,使其不能再作为供水水源。由于破坏了硅藻在春季的生长过程,硅藻的净化作用不能得到实现,反而使得水体变成了一个污染藻类过度生长的碱性污水沟。

硅藻和绿藻的一个重要不同点是鱼类只吃硅藻,而不食用绿藻。硅藻消耗的营养物质可以包埋在沉积物中而去除,而绿藻消耗的营养物质却又回到了水体中。

来自污水处理过程的亚硝酸盐

污水处理厂是另一个饮用水中亚硝酸盐的主要来源。
水中的污染物有四种,即无机颗粒、有机颗粒、微生物和溶解性物质。污水处理厂用格栅、砂滤、沉淀池来去除颗粒物质。这样的物理处理的应用比率比较低,效果也不好。目前的处理方式主要是用生物降解的方法来去除有机污染物。有机物质通过细菌分解,使其中的 C 变成 CO_2,同时会有少量 N 和 P 被释放到水体中。在还原环境下的生物降解过程会产生无害的氮气,但是这个过程很缓慢,费用也比较高,所以一般都会采用氧化生物降解的方法进行污水处理。污水处理厂的专家应该都知道亚硝酸盐的危害,但是他们也不大关注在处理后的水中的亚硝酸盐可能带来的后果。

政府限制污水处理厂的"中水"的回用,但是管理者很少知道需要减少亚硝酸盐的危害。最大亚硝酸盐允许浓度目前是没有被列为污水厂的限值项目中的,以至于很少有中国的污水处理厂会去分析水中的亚硝酸盐。我们的一些分析显示,污水处理厂处理后,水中的亚硝酸盐浓度并没有明显的降低。

由于管理层对于亚硝酸盐危害的无知,污水处理厂的管理者们反而强烈地推荐中水用于灌溉。他们可能不知道在灌溉水蒸发后,亚硝酸盐会留在土壤中。研究表明,使用回用的处理水浇灌的土壤中的亚硝酸盐的浓度是用清洁水浇灌的土壤的 2 倍。留在土壤中的亚硝酸盐在下一次雨水到来时会污染河流或者进入地下从而污染地下水。

过滤型水转换器和自净塘

为解决目前氮的危机,我们已经做了如下的一些工作:

(1) 发明了一种具有很小筛孔和很大的过流速率的

过滤装置,可以有效地去除水中的颗粒物质。所以这样的话,生物降解的方法可以不再使用;

(2) 发明了一种去除过滤后的水中的溶解性的亚硝酸盐的方法;

(3) 发明了一种方法,可以改变自然环境,使得过滤出水不会再发生绿藻等藻类污染问题。

许靖华发明了一种过滤型水转换器,可以用于去除污水中的颗粒物质、无机物质、有机物质以及微生物。为了实现高效的过滤去除,过滤介质的上层由中度大小的细砂构成。为了实现大的过流速度,水转换器还设置了一个打孔管埋在砂砾层的底部。过滤介质和水转换器用一个起保护作用的覆盖物(保护层)隔开,以防止细小的过滤介质进入孔径大的过滤层中。

打孔管与泵相连,通过泵产生一个砂石上水流的负水势。尽管从水转换器到过滤层的水势梯度较小,但是在装置底部和过滤层顶部之间产生的水势是非常大的。水势梯度自身的调控功能保证了水流的持续流出。

利用这样的一个装置,水流通过 $10\mu m$ 孔径的过滤层的流速可以达到 $10\ m^3/m^2$ 之快。细砂层不是主要的过滤层,这一层的水流流速有 $25\ m^3/m^2$。经过处理后的出水非常干净,没有悬浮物质,除了 N 和 P 的指标外,其他指标基本上可以达到回用水 I 级标准(见下页表)。

不同水处理样品的营养成分比较

	A	B	C
Total N	18.89	2.08	1.24
NH_3 N	1.60	0.12	0.11
Nitrate-N	9.85	0.1	0.0032
Nitrite-N	0.15	0.06	0.0004
Total P	0.23	0.068	0.346

样品A是北京一家污水处理厂的水质分析结果；

样品B是经过水转换器处理后的出水的水质分析结果；

样品C是在自净塘中停留了两个月后的水样的分析结果。

  过滤是一个物理过程。一些吸附在悬浮颗粒物上的离子可以通过过滤的方法去除，但是去除污染物也需要进行一些化学或者生物化学的方法。目前有三种可行的除去亚硝酸盐的方法：蒸馏法，离子交换法，反渗透法。在这三种方法中，目前，在北京各大超市出售的"瓶装饮用纯净水"一般是用反渗透法生产的。但是，这个膜处理的过程非常昂贵，用于供水脱氮在经济上的可行性不大。

  在上述对亚硝酸盐的来源的讨论中，我们已经指出，自然水体中的浮游生物能有效地去除水体中的营养物质。藻类通过光合作用利用 NH_4^+ 和 PO_4^{3-} 制造自身生长需要的物质，在这个过程中产生 CO_2 和水。

$$106CO_2 + 236H_2O + 16NH_4^+ + HPO_4^{2-}$$
$$= C_{106}H_{181}O_{45}P + 118O_2 + 171H_2O + 14H^+$$

在上述过程中消耗的 NH_4^+ 由下面的反硝化脱氮作用补充：

$$NO_2^- + 5H^+ + 3Cl^- + H_2O = NH_4^+ + 3HClO^-$$

(在酸性环境中)

$$NO_2^- + 3Cl^- + 3H_2O = NH_4^+ + 2OH^- + 3ClO^-$$

(在碱性环境中)

所以,浮游生物(如酸性环境中的硅藻和碱性环境中的绿藻)都能将自然水体中的亚硝酸盐去除。但是,为什么绿藻还是被视为产生环境污染的有害藻类呢?

在前述中我已经指出,主要引起污染的不是藻类的生长,而是藻类的死亡引起的污染。绿藻有一些藻毒性,所以鱼类也不会食用绿藻,这样冬天绿藻死亡后,它们的残骸就会沉降到水底,有机氮被转换为氨、硝酸盐和亚硝酸盐。一些分解物会产生臭气,而硝酸盐和亚硝酸盐则对健康有着很大的危害。

鱼类是以硅藻为食的,我们已经成功地在自净塘中培养出了硅藻,并能够用这些硅藻养殖一定数量的鱼虾。在大概一个月后,过滤水中的硝酸盐和亚硝酸盐都被去除了(见上页表),水中的亚硝酸盐浓度小于 0.004 mg/L(以 NO_2^- 中的 N 计)。这种方法的脱氮作用是在自净塘中进行的,所以生产这种水的费用也是很低的。

我们知道,硅藻通过光合作用利用 CO_2 和 N、P,当水中的亚硝酸盐被去除后,水体的碱度将会有所增加。为了防治水体再被绿藻污染,我们往自净塘水体中不断地灌入 CO_2,使得水体保持一个微酸的环境,这样就能够抑制绿藻的生长,促进了硅藻的生长。

为了抑制水中的绿藻的生长,水的 pH 应该不超过 7。当水体的 pH 低于 5.8 时,所有的绿藻都会死亡。当绿藻死亡后会有残骸留在水体中,所以还是需要通过过滤作用去除,否则这些残骸分解释放的亚硝酸盐还是会污染水体。

CO_2 的来源可以是 CO_2 储气罐,可以是酸解矿石后产生的 CO_2,可以是用管道系统收集的使用石油染料的工厂排放的 CO_2 废气。利用排放的 CO_2 废气来处理污水,这同时还能减少温室气体的释放,缓解全球变暖趋势。

结　论

我们的技术结合了水转换器的过滤作用和自净塘的脱氮作用,是一种能够实现污水回用的方法。处理后的出水能够直接用于人们生活,也可以回灌地下水。如果这种处理污水和回灌地下水的方式能被采用,那么就不会再有环境污染,不会再有水资源短缺问题,荒漠也能变成绿洲。

3 A Symbiogenetic Origin of Cancer and the Nitrite Paradox

Abstract

Clinical tests in Europe and North America have led to the conclusion that nitrite-rich drinking water should not cause cancer. On the other hand, statistical evidence in China has correlated cancer mortality-rate to nitrite in drinking water. Furthermore, the Chinese cancer-epidemics are brought under control where the source of drinking water is changed to one depleted in nitrite. The apparent contradiction between the West and East can be explained within the theoretical frame work of our theory of symbiogenetic evolutionary origin of cancer. We recognized that cancer cells are genetically different from normal cells, and their paths of metabolism are different. They are however dormant in stem cells, until they are exposed at locations where nitrite is easily accessible to sustain their growth. In other words, nitrite in drinking water does not cause cancer, but their availability encourages cancer growth. The theory can be tested whether cancer prevention or cure can be effected when one drinks nitrite-depleted water; one can monitor the cancer-mortality rate at communities supplied with nitrite-depleted drinking water. While the scientific debate is being resolved, the public should be advised to drink bottled water, which has, in China, a MCL of 0.002 mg/L nitrite-nitrogen.

The Nitrite Paradox

Nitrate is not harmful to human, but digestive processes in human body can transform nitrate into nitrite. Nitrite could cause blood to lose oxygen, and excessive dose is the cause of a rare infant disease methemoglobinemia. That nitrite could cause cancer was once suggested to explain the sharp decline of gastric cancer during the first half of the 20th century; the decline was correlated to the widespread use of refrigeration instead of meat-preservation by nitrite. The various risk aspects of nitrite in food and drink were thus thoroughly investigated in Europe and North America during the last decades of the 20th Century.

Data are irrefutable that nitrites, when ingested in high concentrations, cause methemoglobinemia. The scientific evidence is also clear that N-nitroso compound (NOC) have carcinogenic effects in animal models. In fact, elevated concentration of NOCs has been shown in clinical tests in China to cause cancer in humans as well. However, neither experiments or theories are able confirm a direct link between the presence of nitrites in food and drink and the causation of human cancer. Studies with mice until 1981 were evaluated by Life Science Council of the National Academy of Sciences of the United States, none indicated any carcinogenic effect of nitrite. Later studies until 1993 were summarized by the Committee of Experts on Health Control of Foodstuffs of the Council of Europe. In one experiment mice received drinking water containing 0, 1000, 2500 and 5000 mg/L sodium nitrite for 18 months, no changes in tumor incidences were observed. In another rats receiving drinking water concentration of 0, 0.125%, or 0.25% sodium nitrite for 2 years, no carcinogenic effects were observed. Nitrite is, therefore, officially declared not carcinogenic. The EPA identified a no-observed-adverse-effect level (NOAEL) of 10 mg/L of nitrogen in nitrate, and for 1 mg/L of nitrogen in nitrite. Those guidelines for nitrate and nitrite in drinking water were promulgated to protect infants from developing methemoglobinemia. The EU authorities shared the American judgment that nitrite is not carcinogenic, and the MCL of nitrite drinking water for several European countries, including UK, was in fact raised in 2002 from 0.1 mg/L in to the EU standard 0.5 mg/L of nitrite-nitrogen.

Since the rapid industrial development of China in 1980, there have been remarkable increases of cancer-mortality rate. In west China, west of 110 E. longitude, the rate is less than 100 cancer-deaths per 100000 population per year, with a minimum of 12 in under-developed counties. In the more developed east China, east of 110 E. longitude, the rate is mostly between 100-200 cancer deaths per 100000 population each year, with a maximum of more than 200, which is the about average rate in the West.

The cancer mortality rate in China is related to nitrite concentration in drinking water. The rate (10-20cancer deaths per 100000 population) is low in rural communities where the nitrite concentration is very low ($<$0.005 mg/L nitrite-nitrogen). In "cancer villages" and in many cities, the mortality rate (100-200 persons per 100000 population) and the nitrite concentration (0.1-1.0 mg/L nitrite-nitrogen) are both high, yet they still fall within the average range of the West. It is noteworthy that cancer-mortality rate in China is increasing at communities where nitrite in drinking water increases. Even more significant is the fact that cancer mortality is drastically reduced where the source of drinking water is changed to be nitrite-depleted; lives were saved!

In view of the convincing new evidence, Chinese experts met in 2005 to discuss the nitrite hazard. The government faced a dilemma: A very stringent standard would deprive many cities of their water-supply, but a tolerant limit, in view of available evidence in China, would be irresponsible. Unable to decide, the health officials encouraged the public to consume "purified drinking water," and set a PMCL in 1998 for bottled water at 0.002 mg/L nitrite-nitrogen, $1/500^{th}$ that of the US NOAEL.

The nitrite paradox seems to reflect a difference in popular attitudes toward science. The general public of the West, in the tradition of rational enlightenment, tend to accept the judgment of establishment majority; nitrite is harmless and cancer-deaths are unavoidable. The people of the Orient, having suffered more under human follies, are be more distrustful: the scientific interpretations could be wrong. Furthermore they do not see the need of martyrdom: If there is a way to reduce cancer mortality, such as a change of the source of drinking-water, why should one not try, especially it is not very costly. This article is an attempt to encourage people to save lives, including perhaps that of their own; we offer a scientific explanation of the nitrite paradox, and propose that the public be supplied with nitrite-depleted water even before the scientific issue on the origin of cancer, is completely understood, and the debate resolved.

Cancer is a Genetic Disease

The idea that cancer is an infection by a microbe is not generally accepted. Although cancer cells cultured after the death of a cancer patient can replicate indefinitely and are considered a micro species distinguished by its non-human chromosome number, still the species Helacyton gartneri is not known to occur in Nature, and cannot be cited as evidence of carcinogenesis by microbes.

The current paradigm recognizes that cancer is a genetic disease. There is little understanding, however, as to the origin of the genes. Two recent developments suggested that cancer cells are inherited. In that case, nitrite in drinking water could not have caused cancer, but its presence in blood streams could sustain the growth of cancer cells that already exist.

Cancer, Nitrite and the "Warburg Effect"

Metabolism encompasses all the chemical reactions occurring in living organism. Included are all the molecular processes by which organisms acquire and utilize the energy they need for their growth, maintenance of life, and performance of biologic functions. The entire animal kingdom employs essentially one mode of metabolism-heterotrophy: the use of oxygen to respire organic food molecules. There are two main ways to generate energy in human cells in the form of adenosine triphosphate (ATP): oxidative phosphorylation and glycolysis. Normal cells rely overwhelmingly on the path way the former; oxygen in blood stream is taken up by the mitochondria organelles in cells, 38 moles of ATP are derived from the oxidation of 1 mole of glucose, yielding a vast amount of energy. The metabolic mode switches to glycolysis in the cytoplasm at times of oxygen deprivation. Glycolysis is a less effective metabolism, producing only two ATP molecules per glucose molecule.

In contrast to animals, single-celled organisms have more metabolic modes: the bacteria have more than 20. One such mode is anaerobic respiration. It is known that denitrification, the reduction of nitrate or nitrite to nitrous oxide or nitrogen, is the major mechanism which fixed nitrogen returns to the atmosphere from soil and water. This denitrifying ability has been found in many microorganisms belonging to numerous groups of bacteria and Archaea. The genes encoding the denitrifying reductases have been studied in a few species. Bacteria of the genus Desulforvibrio uses nitrite respiration, and the nitrite reductase is localized in the cytoplastmic membrane of D. gigas.

The use of a specific compound such as nitrite other than O_2 as the source of oxygen is of particular interest to our inquiry, because of the so-called "Warburg effect" of cancer cells. Otto Warburg, the only person to win two Nobel Prices in Medicine, proposed that cancer has only one prime cause: cancer is caused by defects in oxidative phosphorilation. Instead of respiration in the mitochondria, the cancer cell is reverted to the more primitive form of energy generation by the path of glycolysis. In his 1927 paper, he specifically mentioned the role of nitric oxide in depriving a cell of oxygen.

Warburg cancer theory fell on deaf ears; the establishment majority conceded that altered metabolism could be a symptom of cancer, but not the primary cause. Warburg went to his grave in 1970 insisting he was right, but for 30 years his cancer theory was buried along with its originator. Warburg's theory was only exhumed in 2004 when Chi Dang and Craig Thompson both found that glycosis in cancer cell, or a shift in energy production - the so-called Warburg Effect, is not a by-product of cell's transformation into cancer, but a requirement. Genes in normal cells programmed to make mitochondria were replaced by genes in cancer cells to synthesize DNA and to make the cell proliferate.

The use of the term shift to denote the "Warburg Effect" is subjective, and implies a transformation, a change from normal cell to cancer cell. The orthodox Neo-Darwinism teaches that such transformation could only be a consequence of gene mutations: There has been a shift of mechanism of metabolism, but the normal cells has been transformed into cancer cells. An alternative is to assume that the normal cells have ceased their functions while dormant cancer cells take over the "shift" and perform metabolic duties.

The orthodox "clonal evolution" postulated that tumors arise from normal parental cells that mutate and generate abnormal offsprings. The alternative requires no mutations: Perhaps the cancer cells have always been there, but they are dormant until the opportunity for anaerobic metabolism presents itself. As Warbug indicated, the growth can be stopped when cancer cells are killed or when they are starved. Oxygen can be toxic to cancer cells, and ozone has been proposed as a cancer drug, but nitrite is not toxic either to normal or to cancer cells, but nitrite-glycosis requires a continued supply of mineral-oxygen. Nitrite in drinking water has not caused cancer, but tumors, initiated and driven by a single abnormal adult stem cell, requires nitrite to sustain their growth.

Biology of Cancer Stem Cells

In a cancer cell, the mutation theory postulates that several genes change (mutate), and such cells are considered defective. Defective cells would invade the normal tissue where cancer starts, or to travel in the bloodstream (metastasize) to remote parts of the body, where it continues to divide. The theory believes that cells become cancerous when part of a gene is lost (deleted), when a part of a chromosome is rearranged (translocation), or when a defect occurs in the DNA. The recent development of stem-cell science is, however, revolutionizing the field of cancer biology, and has provided us with new perspectives on the origin of cancer.

The cell is the fundamental unit of life. It is the smallest structure of the body capable of performing all the processes that define life. Each of the organs of the body, such as the blood, breast, brain, etc. consists of specialized cells that carry out the organ's functions. To assure the proper performance of each organ, worn out or injured cells must be replaced, and particular types of cells must increase in response. For example, the bone marrow can increase the production of oxygen-carrying red blood cells in response to bleeding. Similarly, the liver or endocrine organs frequently respond to injury by regenerating damaged cells. Cells grow by division, and the growth is controlled by DNA. Cells divide only when they receive the proper signals. For example, if a person loses blood, a growth factor called erythropoietin which is produced in kidneys circulates in the bloodstream and tells the bone marrow to manufacture more blood cells. It seems that cancer only arise only in injured organs and tissues that display lost or diminished regenerative ability.

The stem-cell theory postulates that the growth of cancers, leukemia, pancreatic, breast and brain cancers, etc. is to be linked to long living cells in the body and they possess properties in common with undifferentiated embryonic cells. Stem cells respond to signals to regenerate damaged cells. Not all, but only a very few of the stem cells are cancerous, and they are able to reproduce themselves and to differentiate into other tumor cells.

Helacyton gartneri

Stem cell biology has come of age. Unequivocal proof that cancer stem cells exist, their properties and genetic programs are begun to be delineated. Cancer cells are different from normal cells by their indefinite potential for self-renewal that drive tumorigenesis.

When human body cells are put into a cell culture,

they weaken and die quickly, usually within about 50 divisions. Without the rest of the support structure a heart, blood circulating, a digestive system, etc. body cells cannot survive. Body cells also age. The basic mortality of the cells reflect the mortality of the organism they comprise. Cancer cells are, however, to an exception. After Henritta Lacks died of cancer in 1951, George Gey, head of the tissue culture research at Johns Hopkins, discovered that cells from Henritta's tumor would not only survive and multiply outside of her body, but they did not age either. In fact, they are still alive, and live independently of the body from which they came. The HeLa cells reproduce, they consume, excrete, and they do everything an independent living organism usually does.

It has been suggested that horizontal gene transfer from human papillonmarvirus 18 (HPV18) created the He(rietta)La(cks) genome. However, the cancer genome is different from either parent genome in various ways. HeLa cells, for example, have a modal chromosome number of 82. Leigh Valen controversially described them as an example of the contemporary creation of a new species Helacyton gartneri, Van Valen & Maiorana. There are many strains of Hela cells as they continue to evolve by being grown in cell cultures. All HeLa cells may have been derived from the original cells from Lacks. Alternatively a substantial fraction of in vitro cell lines are actually Helacyton gartneri, but their original cells have been overwhelmed by the rapidly growing population derived from HeLa contaminant cells.

If cancer cells belonging to free-living species, their presence in human body could suggest a parasitic or symbiotic relation. Yet HeLa is a new species, no known microbe species could be identified as the invading parasite or symbiont. Van Valen identified, therefore, HeLa as a product of symbiogenetic evolution. With this theory, one can postulate that the abnormality of cancer cells has evolved symbiogenetically. They are inherited and are present already in the stem cells, even before they are called upon to repair cell damages. The cancerous cells are abnormal because they do not have mitochondria or their mitochondria organelles have been too badly damaged, to produce energy by taking up oxygen in blood stream for oxidative phosphorilation. Alternatively, the cancer cells have acquired genomes from symbiotic cells which have the mode of anaerobic metabolism.

Evolution of Cancer Cells through Symbiogenesis

Endosymbiosis and Symbiogenesis

The minimum unit of life is the cell. Genomes are embedded deep in cells. Bacteria are unicellular, where plants and animals are multi-cellular. The minimum genome of a free-living bacterium has about 500 genes whereas the genomes of plants and animals can be one hundred times as large. The Neo-Darwinian theory of evolution teaches the paradigm of variations by genetic mutation, adaptation, and natural selection. A small minority of biologists have disagreed. In 1927, the same year when Otto Warburg discovered the dysfunction of mitochondria in cancer cells, Ivan Emmanuel Wallin, a professor of anatomy, wrote Symbionticism and the Origin of Species to postulate that prokaryotes acquired their mode of aerobic mode of metabolism through the acquisition of the mitochondria genome during symbiontic evolution. Wallin the rebel did not know of work by the Konstantin Merezhkovsy or Andrey Famintsin. He did not know that acquisition and inheritance of microbial symbionts had been called symbiogenesis. He did not know that the Europeans had presented the same thesis: symbioses and their consequences generate evolutionary changes.

Supported by the discoveries in molecular biology and paleontology, Lynn Margulis has recently revived the theory of symbiogenetic evolution through acquisition of genomes. She considered Kwang Joen's experiment showing amoeba, infected by and acquiring the genome of another bacterium, a laboratory example of the origin of a new species.

Microbes living on their own tend to merge with other forms of life and become symbiotic. The one symbiotic cell is enclosed within another, the first step is taken toward symbiogenesis. This endosymbiosis is best illustrated by amoebas. Those single cell organisms appear to be not much more than cytoplasm held together by a flexible cell wall. Floating in this cytoplasm are, however, all kinds of cell bodies. The most obious is the nucleus. Some species have one nucleus, but others have many nuclei. Apart from those, the cell may contain all kinds of inclusions. Flurescent bacteria, for example, were demonstrated to be abundantly spread as single cells, throughout the cytopolasm of the giant ameba Pelomyxa palustris. The endosymbiots were identified by their cell biology as methanogenic bacteria; so that individuals of P. palustris are known to have produced methane over long period of incubation. Pelumyxa cells have no mitochondria, and the energy is supplied for cell use through glucolysis by organelles, and the source of energy is the methane produced by symbiotic methanogenic bacteria.

Evolution by Symbiogenesis

Life began in Archaen some 4 billion years ago. Organisms with bacterial cell organization have no nuclei. Not classifiable into distinct species, they all belong to a single worldwide species. Speciation is a property only of nucleated organisms, and it began with the earliest protoctists during early Proterozoic, some 2.5 billion years ago. Species originated by symbiogenesis, when microbes and larger life forms fuse permanently. The fusion, or the acquisition of the genomes of a formerly symbiotic species, is the symbiogenesis theory of the origin of species. By now most scientists involved in biology or paleontology accept that once revolutionary idea.

The prokaryote cells, with membrane-bound neclei, have evolved from symbiotic assemblages. Eukaryotic parts such as mitochondria, cilia, and microtubules evolved as free-living species before their fusion created new species through the incorporation of bodies and genomes. Fusions of eukaryotic cells are to produce fungi, plants, and animals, the other three members of the Five

Kingdoms of living organisms.

Bacteria had evolved nearly all the important metabolic traits before the appearance of eukaryotes. Their metabolic modes of bacteria are a reflection of the history of the Earth's atmosphere. The primeval atmosphere consisted of carbon dioxide and water, as well as other volcanic emanations. With the beginning of carbon cycling by bacteria, Earth escaped the fate of Venus of overheating, when anaerobic bacteria evolved 3.8 billion years ago and their fossilization as mineral carbon removed CO_2 from the atmosphere. The depletion of carbon dioxide may have been carried to the extreme on Mars, where anaerobic bacteria may once have existed. The extreme caused the freezing of CO_2 in Martian atmosphere into dry ice, but the Earth escaped the fate of Mars, when methanogenic bacteria evolved some 3 billion years ago. The conversion of mineral carbon to methane provided the necessary greenhouse to keep the Earth warm at that distant time of reduced solar radiation:

$$2C + 2H_2O = CH_4 + CO_2$$

The continued pollution of methane was, however, again to cause over-heating. Gaia, the metaphor for a self-adjusting terrestrial system, stepped in again. Isotopic evidence indicates the gradual decline, some 2.5

billion years ago, of the methanogenic bacteria. They were replaced by the methylotrophic and cyanobacteria. The former takes mineral oxygen from nitrite to burn methane in order to produce energy for their cells:

$$CH_4 + 2NO_2^- = CO_2 + 2H_2O + 2NO^-$$
$$2NO^- + 4H^+ = 2H_2O + N_2$$

The denitrification by methylotrophic bacteria gave the nitrogen in atmosphere.

At the same time photosynthesis by cyanobacteria produced the oxygen.

The major change 2.5 billion years ago, or the Oxygen Revolution in Earth History, marks the transition from the Archaen to the Proterozoic Era. Cyanobacteria became the dominant organism of the Preoterozoic, extracting carbon dioxide from the atmosphere to produce oxygen. The reduced greenhouse effect led finally to global glaciations during the time of Snowball Earth just before the beginning of the Cambrian Period. The history of the life on Earth is then a record of alternating carbon-dioxide producing and carbon-dioxide absorbing organisms, resulting in periods of extreme glaciation (some 300 million years ago at the end of the Carboniferous) and of intense global warming (some 150 million years during the late Jurassic).

The history of the life on Earth is also a history of

symbiogenesis. Chloroplast in plants, the ability to synthesize with light energy carbon dioxide and water, was derived from once free-living photosynthetic bacteria: the genome was acquired by plants. Mitochondrion, the ability to use oxygen to respire organic food molecules, was derived from once-free living bacteria similar to bdellovibrio: the genome was acquired by plants and animals. Within the framework of the evolutionary history, we could place the beginning of symbiogenesis of mitochondria with the cells of animals and plants at the time Oxygen Revolution. Oxygen was poison to anaerobic bacteria, and the appearance of an oxygen-atmosphere was to wipe out almost all the once-predominate methanogenic bacteria, except the rare survivors in anaerobic environments. The plentiful supply of origin led to the evolution of plants and animals which obtain their energy through mitochondria respiration.

The survivors in anaerobic environments had to seek new symbiotic partner.

The amoeda Pelobiontida. are eukaryotes, but they are one of the few groups of eukaryotes that have not acquired mitochondria. Those primitive amitochondriteates arose after the diversification of the earliest evolved protoctists, and they evolved through the acquisition of symbiogenetic genomes sequences that are common to eubacterial, archaeal, and eukaryotic cells. The Pelobiontida have not acquired mitochondria, because the group diverged before the endosymbiotic origin of mitochondria. Those organisms have symbionts specialized in anaerobic metabolism developed before the appearance of oxygen in atmosphere.

  The cancer cells may have acquired their metabolic mode like the giant amoeba. Free living eukaryotes acquired the genomes from Desulforvibrio sp. or from some other anaerobic bacteria which utilized nitrite for glucolysis. Cancer cells like the giant amoeba also evolved when there was no oxygen in the atmosphere some before symbiogenesis produced mitochondria in plants and animals. When plants and animals evolved, the cells fused together are not only those with mitochondria, they also acquired the genomes of cancer which are to produce cancer stem cells in human embryos. They are however dormant. They only again become active when the microchemical environment within human (or animal) body again becomes similar to the time of anoxic atmosphere which prevailed more than 2.5 billion years ago. Cells with anaerobic mode of metabolism take over, and tumors grow. Their continued growth depends, however, upon a continuous supply of nitrite. Within such a theoretical framework, nitrite is not carcinogenic in causing the growth of cancer; but it is "carcinogenic" in the sense that nitrite is the nutrient necessary to sustain the growth of cancer. When there is no nitrite in drinking water, the supply is cut off, and cancer cells are starved to death. There should be remission of cancer growth.

Nitrite Sustains Cancer Growth

Academician Lu, former head of China's Cancer Research Institute, went to Linxian (Henan) where a cancer epidemic was linked to high nitrite concentration in drinking water. Lu and others conducted clinical experiments and found for the first time that human esophageal carcinoma could be induced by N-nitrosamine, they reported in 1989:

In the present study, human fetal esophageal epithelium was cultured for 3 weeks with NMBzA obtained in Linxian County (a high incidence area of esophageal cancer). Then the explants were heterotransplanted to mesentery of BALB/c nude mice which were continually fed with the nitrosamine in drinking water for 8 months. The results showed that in NMBzA-treated mice, a small tumor was found on the mesentery two months after transplantation. The small tumor grew to 2.18 cm in size 8 months after transplantation. No tumor was observed in control nude mice… Pathology of the induced tumor showed squamous cell carcinoma. DNA extracted from the tumor induced by NMBzA was hybridized with Alu sequence···indicating that the tumor is of human origin.

The Linxian Experiment was a difficult experiment: Academician tried on more than 300 fetuses, before he succeeded in culture cancer stem cells in N-nitroso solution. The number of cancer cells must have been extremely rare. The experiments the experiments suggest cancer stem cell was activated and esophageal carcinoma was induced in only one of three hundred experiments in N-nitroso solution. The role of nitrite-in-drinking water was actually told me personally by Academician Lu that the tumor in mice only grew after they were fed with nitrite-solution for 8 months after the transplation of the cancerous tissue inside the mice. Strictly speaking, cancer is not caused, but the growth of cancer is sustained, by the nitrite in drinking water.

The essence of our hypothesis is to postulate that cancer cells take nitrite-oxygen, whereas normal cells with mitochondrion organelles take oxygen in blood stream to convert glucose into energy. The role of nitrite is important to the cancer cells as the role of oxygen to normal cells. Normal cells in our body may not be destroyed by nitrite in our blood stream; they only die when there is no oxygen. Nitrite is not poison, and an over-supply of nitrite to blood stream may, in the extreme case, cause methemoglobinemia, which has been observed in some clinical tests. Nitrite is, however, not going to cause normal cells to become cancerous. Therefore, no amount of nitrite intake by experimental animals is to induce cancer, as scientists from Europe and America have concluded.

Nitrite is as necessary to cancer cells as oxygen is for normal cells. The growth of cancer cells has to be sustained by continuation of adequate supply of nitrite. When the source of drinking water is nitrite-depleted, cancer cells starved, and cancer goes into remission. Such a mechanism explains why the cancer-mortality rate is high in Chinese communities where nitrite in drinking water is high, and why a change of the source of drinking water has successfully prevented the further spread of cancer epidemics.

Academician Lu also expressed an opinion why smoking causes cancer. Nitrite in tobacco combines with nicotine to form nitrosamines, which could be cancerogenic. Later, when cancer stem cells are activated, nitrite is continually supplied by habitual smokers to favor the growth of cancer cells.

Cancer Prevention and Cures

On cancer prevention and cures, we refer the readers to the work of Warburg that cancer responds to oxygen-deprived condition; he suggest the supply of oxygen, such as the use of ozone as a clinical drug. We suggest a cheaper and more natural prevention and cure; drink purified nitrite-fee water.

The Hsu-Matsumoto theory of origin of cancer cells through acquisition of genomes in evolution can be tested. We are contacting colleagues in different parts of the World to make a systematic evaluation of cancer-mortality rate to nitrite-indrinking water. We are to introduce nitrite-depleted drinking water to "cancer villages" to observe if the cancer mortality rate is reduced proportionally. We are to cooperate with cancer-biologists to study the cancer cells to search for organelles in cancer cells which may have been acquired symbiogenetically from Desulforvibrio or methylotrophic bacteria.

Cancer occurrence and mortality-rate in China are drastically reduced where nitrite-depleted drinking water is supplied to the public. We personally do not see the need that the public in Europe and North America should suffer while their experts are dedicated to win their scientific debates. There should be no ill consequences to drink purified clean water, even if our theory is proved incorrect. One may not have to change his life style. He could still have his bacon and eggs for breakfast, even though cured meat has a very high nitrite content. If the theory is correct that cancer cells are in all of us, but tumors would not grow if they are not sustained by a continuous supply of nitrite. The cancer cells can remain dormant when they are, as when they are hidden in the bone marrows, deprived of nitrite as nutrient for their growth. Only when cancer stem cells are dislocated to locations, such as to blood, to organs of lungs, digestive system, etc. , where the supply of nitrite is easily accessible, do we need to take precaution. Finally a patient can still try to starve the cancer cells, if our theory is proven correct; he could start to drink nitrite-depleted water after cancer is diagnosed.

References Cited

1. Paik , D. C. , Saborio, D. V. , Oropeza, R. , and H. P. Freeman, 2001. The epidemiological enigma of gastric cancer rates in the US . Int. Jour. Epidemiology,30:181-182.
2. AMA Council of Scientific Affairs, 2004. Report 9 (A-04). American Medical Association,pp. 9.
3. Lu, S. X. , Chui, S. X. , Yang, W. X. , Hu, X. N. , Guo, L. P. , and Li, F. M. , 1991. Relevance of N-nitrosamines to oesophageal cancer in China , in I. K. O'Neil et al. (editors) Relevance to Human Cancer of N-Nitrose Compounds, Lyon : International Agency for Research on Cancer, pp. 11-17. (Revised English translation of a 1989 article first published in Chinese) in Chinse Cancer Research, 11:401-403).

4. National Academy of Sciences-National Research Council-Life Sciences Council, 1981. The Health effect of nitrate, nitrite and N-Nitroso. Washington D. C. : National Academy of Sciences. Washington D. C. @ National academy Press.
5. Committee of Experts on Health Control of Foodstuffs, 1993. Nitrates and Nitrites in Foodstuffs. Strasbourg : Council of Europe Press, 1993, pp. 128.
6. Lai, D. , 2004. Geostatistical analysis of Chinese Cancer mortality. Jour. Data Sci. 2: 177-193.
7. Hsu, K. J. , Ye, W. , Kong, Y. , Li, Dong, Hu, F. , 2007. Use of Hydrotransistor and De-nitrification Reservoir to Produce Purified non-Carcinogenic Water for Drinking and for Groundwater Recharge. Proc. NAS (in press).
8. Van Valen, L. M. and Maiorana, V. C. , 1991. HeLa, a new microbial species. Evolutionary Theory 10:71-74.
9. Philipport, L. , 2002. Denitrifying genes in bacterial and archaeal genomes. Biochim biophys Acta, 1577 (3): 355-76.
10. Barton, L. L. , LeGall, J. , Odom, J. M. , Peck, H. D. , 1983. Energy coupling to nitrite respiration in the sulfate-reducing bacterium Desulforvibria gigas. Jour. Bacteriology for 1983, pp. 867-871.
11. Warburg, O. , 1927. Ueber did Wirkung von Kohlendioxyd und Stick-Oxyd auf Atmung und Garung. Biochmie Z. 189: 354.
12. Garber, K. , 2004. Energy boost: the Warburg Effect returns in a new theory of cancer. Jour. National Cancer Institute, 94:1805-1806.
13. Ruggiero, R. A. and Bustuoabad, O. D. , 2006. The biologi-

cal sense of cancer: a hypothesis. Theoretical Biology and Medical Modelling, 3:43.
14. Masters, John R., 2002. HeLa Cells 50 years on: the good, the bad and the ugly. Nature Reviews Cancer 2:315-319.
15. Van Valen, L. M., and Maiorana, V. C., 1991. HeLa, a new microbial species. Evolutionary Theory 10:71-74.
16. Wallin, J. E., 1927. Symbionticism and the Origin of Species. Baltimore :William and Wilkins.
17. Margulis, Lynn, 1997. Microcosmos: Four billion years of microbial evolution. Berkeley:University of California Press.
18. Van Bruggen, J. J. A., Stumm, C. K., Vogels, and G. D., Symbiosis of methanogenic bacteria and sapropelic protozoa. Archives Microbiology, 136:89-95.
19 Cavalier-Smith, T., 2002. The phagotrophic origin of eukaryotes and phylogenetic classification of Protozoa. Int. Jour. Systematic and Evolutionary Microbiology, 52:297-354.
20. Hsu, K. J., 1993. Gaia has a strange attractor-Interactions of geosphere and biosphere during 4-billion years of Earth history. Proc. Int. Symposium.
Taichung, Taiwan: National Museum of Natural Science Press, pp. 117-126.
21. Hsu, K. J., 1990. Is Gaia endothermic? Geol. Mag. 129: 129-141.
22. Marguslis, Lynn and Sagan Dorion, 2002. Acquiring Genomes: A theory of the Origins of Species. New York: Basic Books.
23. Stiller, J. W., Duffield, E. C. S., and Hall, B. C., 1998. Amitochondriate amoebae and the evolution of DNA-depend-

ent RNA polymerase II. Proc. NAS, 95:11769-11774.
24. Lu, S. X. , Chui, S. X. , Shi, J. G. , 1989. Cancer induced by N-nitrosamines in human stem cells. Chinese Cancer Research, 11:401-403.

亚硝酸盐的特点与癌症共生进化的源起

摘　要

　　欧洲和北美洲的临床试验表明,富含亚硝酸盐的饮用水不会引起癌症。但是,中国的统计数据表明,癌症死亡率与饮用水中的亚硝酸盐有着直接的相关性。而且,在中国那些饮用不含亚硝酸盐的饮用水的地区,癌症发生率是可以控制的。东西方之间这样显而易见的矛盾是可以在癌症的共生起源进化学理论框架下予以解释的。癌细胞与正常细胞的基因组成和代谢途径是不一样的。在干细胞中,癌细胞是处于休眠状态的,直到这些细胞接触到了亚硝酸盐。也就是说,饮用水中的亚硝酸盐激活了癌细胞,同时维持其生长。这样的说法可以通过监控一个地区饮用不含亚硝酸盐的水是否能够预防甚至是治疗癌症这样的试验来证明。在中国,瓶装水的亚硝酸盐的限值浓度是 0.002 mg/L。当以上这个科学争论解决后,应该建议和说服人们去饮用瓶装水。

亚硝酸盐悖论

　　硝酸盐对人体是没有害处的,但是人体内的消化过程却能将硝酸盐转化为亚硝酸盐。亚硝酸盐能引起人体血液缺氧,过量的亚硝酸盐还能引起一种较为少见的婴幼儿疾病——高铁血红蛋白症。在20世纪前50年间,人们开始采用冰箱冷藏的方法来保存食物,而不再用含有亚硝酸盐的物质进行保藏,这就使得胃癌发生率降低了很多,这是可以用亚硝酸盐能致癌来说明的。之后,在20世纪的最后十几年里,欧洲和北美洲就开始对使用和饮水中的亚硝酸盐的存在可能造成的各种风险进行评价和调查。

不可反驳的数据表明,摄取了高浓度的亚硝酸盐会导致高铁血红蛋白症的发生。在动物身上的试验表明,含有亚硝基的物质(NOC)对动物有着致癌作用。实际上,中国的临床经历也证明,高浓度的亚硝酸盐对人体也会有致癌作用。但是,还没有实验或者理论能确定食物及饮水中的亚硝酸盐和人类癌症的发生有着直接的联系。直到1981年,由美国国家科学院生命科学委员会进行的在老鼠身上的试验也没有发现亚硝酸盐有着致癌性。1993年,欧洲食品健康控制专家委员会进行了后续的试验研究。在一个试验中,他们用含有0、1000、2500和5000 mg/L的亚硝酸钠的饮用水来喂养老鼠,18个月后老鼠身上的瘤并没有变化。在另外一个试验中,用浓度为0、0.125%或0.25% mg/L的亚硝酸钠的饮用水来喂养老鼠,两年后,老鼠身上也没有发现癌变。所以,一般认为,亚硝酸盐是没有致癌性的。美国国家环保局规定了水中不会产生危害作用的硝酸盐的限值浓度是

10 mg/L(以 N 计),亚硝酸盐的浓度是 1 mg/L(以 N 计)。人们用这些饮用水中的硝酸盐和亚硝酸盐的限值浓度的标准来防止婴幼儿得上高铁血红蛋白症。欧盟也认为亚硝酸盐是不会有致癌作用的,包括英国在内的很多欧洲国家 2002 年将饮用水中的亚硝酸盐的浓度限值从 0.1 mg/L 提高到 0.5 mg/L。

从 1980 年中国的工业迅速发展以来,中国的癌症死亡率有了很大的增加。在中国的西部地区,每年因为癌症死亡的人数少于 100 人/100000 人,在很多区县最少为 12 人/100000 人。而在发达的东部地区,每年因为癌症死亡的人数约为 100~200 人/100000 人,甚至有些地区超过 200 人/100000 人,这与西方发达国家的平均癌症死亡率是差不多的。

中国的癌症死亡率是与饮用水中的亚硝酸盐有着直接的相关的。在农村地区,亚硝酸盐浓度较低(<0.005 mg/L,以 N 计),癌症死亡率也较低(10~20 人/100000 人)。在一些城市的"癌症村"中,他们的癌症死亡率(100~200 人/100000 人)和亚硝酸盐浓度(0.1~1.0 mg/L,以 N 计)都很高,不过这些指标还是低于西方国家的平均水平。值得注意的是,在中国那些饮用水中亚硝酸盐浓度增加的地区,癌症死亡率也在增加;而在那些不再饮用含亚硝酸盐的饮用水的地区,癌症的死亡率却在明显下降。这就意味着,不再饮用含有亚硝酸盐的水源实际上就是在拯救生命!

鉴于这些新的证据,中国专家在 2005 年专门讨论了亚硝酸盐的危害问题。这样,中国政府就陷入了一个两难的境地:制定严格的标准,那么会造成很多城市的供水困难;没有一个严格的标准,又是对人民身体健康不负责任的。不知如何决定和取舍,中国的健康和卫生部门就鼓励人们去购买"纯净水"饮用,同时在1998年将瓶装水的亚硝酸盐浓度限值定在 0.002 mg/L(以 N 计),这是美国规定的最大亚硝酸盐浓度限值的 1/500。

亚硝酸盐悖论似乎反映了公众对于科学的不同态度。西方接受理性文明教导的大众一般都接受这样的说法,认为亚硝酸盐是无害的,而癌症和死亡又是无法避免的。而东方人却比较怀疑科学解释的准确性,而且,他们还觉得,如果有一种方法可以减少癌症的死亡,比如通过改变饮用水源,那么怎么会没有人去尝试这样便宜的方法呢?本文试图鼓励人们珍惜生命。我们在文中对于亚硝酸盐悖论提供了一个科学的解释,并且希望不管癌症的发病原因是不是被彻底确定以及这个悖论和争辩是否已经解决,人们都能饮用到不含亚硝酸盐的饮用水。

癌症是一种由基因决定的疾病

在基因学上,癌症并不被认为是一种由微生物引起的传染性疾病。尽管病人的癌细胞死亡后能够进行不确定的自我复制,同时被认为这是一种与人类染色体数目不一样的的微生物,但是 Helacyton gartneri 这种致癌微生物在自然界中还没有被发现,所以也不能作为微生物引起致癌作用传染病的证据。

目前,人们普遍认为癌症是一种由基因决定的疾病。但是,关于这种基因的起源也是知之甚少。近来有两方面的发展揭示癌症细胞是通过遗传得到的。在上述例子中,饮用水中的亚硝酸盐并不能引起癌症,但是亚硝酸盐在血液中的存在能够维持业已存在的癌细胞的生长。

癌症、亚硝酸盐和"Warburg 效应"

在活的生物中都存在着新陈代谢的化学作用,也包括生物体获取并利用它们能得到的能量进行生长、生命的维持以及其他生命运动的这种分子水平上的反应过程。整个动物王国都进行着一种重要的异养新陈代谢作用:利用氧气消化有机食物分子。人类进行生命活动需要的能量以三磷酸腺苷(ATP)的形式存在,而人类细胞可以通过以下两个途径获取这些能量:氧化磷酸化和醣酵解。正常的细胞主要通过氧化磷酸化的过程来获取能量,细胞中的线粒体利用血液中的氧气,氧化 1 分子的葡萄糖就能产生 38 分子的 ATP,获取大量的能量。在血液中氧气缺乏时,细胞质中就会进行醣酵解的代谢过程。在醣酵解过程中线粒体利用 1 分子葡萄糖只能产生 2 分子的 ATP。

与动物相反,单细胞的生物体能进行多种形式的新陈代谢作用,细菌甚至能进行超过 20 种方式的新陈代谢。其中一种方式是无氧呼吸。研究表明,细菌将硝酸盐和亚硝酸盐通过反硝化的作用代谢成氮气或者一氧化二氮的形式,这实际上是一种最主要的将土壤和水中的氮还原到大气中的方式。这种去氮的能力已经在很多种微生物中发现了,包括种类繁多的细菌和古细菌。人们对某些种类微生物中的反硝化去氮作用的还原酶的基因进行了研究。研究发现,细菌的 Desulforvibrio 基因控制进行亚硝化呼吸作用,而这种进行亚硝化的还原酶是位于 D. gigas 的细胞膜上的。

因为癌细胞的"Warburg 效应"的存在,人类对于这些微生物利用亚硝酸盐而不是氧气进行呼吸的作用非常感兴趣。历史上唯一获得过两次诺贝尔医学奖的人 Otto Warburg 认为,癌症的发生只有一种根本的原因:氧化磷酸化过程中的错误。癌细胞不像正常细胞那样在线粒体中进行呼吸,它会通过醣酵解这种更为简单的方式来获取能量。在 Otto Warburg1927 年发表的文章里,他认为氮的氧化物在夺取细胞的氧气的过程中有着重要的作用。

Warburg 的癌症形成理论并没有引起人们的重视,大多数研究者认为新陈代谢方式的转变只是发生癌症的一个症状,而不是癌症产生的根本原因。Warburg 到 1970 年死为止都一直认为自己是对的,但是之后的 30 年里,这个癌症理论也随着它的创立者一起被埋入了坟墓。2004 年,Chi Dang 和 Craig 都在癌细胞中发现了糖酵解过程,此时 Warburg 的理论才被重新认识,人们才真正意识到微生物获取能量的 Warburg 作用不是细胞癌变的副产物,而是细胞的一种获取能量的方式。正常细胞中编码线粒体的基因被癌细胞的基因取代,并进行 DNA 的合成以实现癌细胞的扩展。

人们主观地用"Warburg效应"这个术语来表达正常细胞向癌细胞的转变,而正统的新达尔文主义认为这种转变只能是基因变异的一种结果:确实存在代谢作用的转换,但是癌细胞是由于正常细胞的基因变异作用导致的。还有一种说法是正常细胞不再执行它们原先的功能,而占主导地位的癌细胞接管了控制权从而执行了代谢的功能。

正统的"物种进化理论"假定瘤是因为正常的母代细胞的变异产生了不正常的后代细胞。还有一种说法认为不存在变异,癌细胞本身就是存在的,但是直到有厌氧代谢作用的发生才会导致它们被激活。根据Warburg的预测,当癌细胞被杀死或者它们处于饥饿状态的时候,癌细胞的生长就会停止。氧气对于癌细胞来说是有生理毒性的,而新鲜的氧气还一直被认为有治疗癌症的作用。亚硝酸盐本身对于正常细胞和癌细胞都没有毒性,但是亚硝酸盐-糖酵解作用需要不断地提供矿化了的氧。饮用水中的亚硝酸盐并没有引发癌症,但是由一个不正常的成熟干细胞长成的瘤细胞却需要亚硝酸盐来维持其进一步生长。

癌干细胞生物学

在一个癌细胞中,基因变异理论认为有一些基因是产生了变异的,而这些细胞是有缺陷的。这些有缺陷的细胞会侵略其他的正常组织,并在那引起癌症,或者通过身体的血液流动到达其他部位,且不断地进行分裂和繁殖。这个理论认为,当细胞的某部分基因丢失(或被删除了)、染色体部分重排或者DNA发生了错误后,该细胞就成为了一个癌细胞。但是干细胞科学的最新研究给癌细胞学领域带来了革命性的改变,给我们提供了对于癌细胞来源的一个全新的认识。

细胞是组成生命体的基本单位,是生物体进行各种生命活动的最小结构单位。每一个人体的组织,如血液、乳房、大脑等等,都是由特殊的能执行这些组织功能的细胞所组成的。为了确保每一个组织都能够正常运行,死亡的或者"伤残"的细胞都会被代替掉,这样就需要有新的该种细胞的生成。比如说,骨髓能促进携氧的血红细胞的产生,这是对组织失血的一种反应。同样的,肝脏或者内分泌的组织也能够产生新的细胞来代替受损的细胞。细胞通过分裂来生长和繁殖,这个过程由DNA控制。只有在接收到正确的信号后,细胞的分裂过程才会开始。举例来说,如果一个人流血了,那么生长因子会通过血液中的红细胞生成素(由肾脏产生)来指挥骨髓制造更多的血液细胞。这样看来,好像癌细胞只会在那些没有了再生能力的器官和组织中才会产生。

干细胞理论认为,癌细胞的生长,如白血病、胰腺癌、乳癌、脑癌等癌症的发生,是与身上的活细胞一起的,它们和与之并无二致的胚胎共同地利用和占有组织资源。干细胞根据信号,对受损的细胞进行修复和再生,不是所有的干细胞都是有致癌作用的,干细胞能够再生,并且不同于瘤细胞。

海拉细胞株

对干细胞已经进行了很多年的研究,确实的证据证明干细胞是存在的,它们的特性和基因学特征也在逐渐被发现。癌细胞与正常细胞的一个最大的不同点是癌细胞能不断地进行自我复制,并最终导致了肿瘤的发生。

当人类正常体细胞在培养基中进行培养时,它们会很快衰竭并死亡,它们往往生存不了 50 次分裂的时间。如果没有支撑其生长的心脏、血液循环或者生殖系统的修整,体细胞就不能生存。当然,体细胞也是会老化的。细胞的死亡反映了它们所组成的器官的死亡。但是癌细胞却有所不同。当 Henritta Lacks 1951 年死于癌症后,Johns Hopkins 的组织培养研究中心主任 George Gey 发现从 Henritta Lacks 的肿瘤中分离的癌细胞不仅能够在体外生存并自我复制,而且它们还不会老化。事实上,它们是仍然存活的,并且是不依靠于它们的寄主组织的。它们能像一个正常的活的器官细胞那样进行自我复制、新陈代谢、排泄以及其他任何生理活动。

一般认为,人体 papillonmarvirus18 基因的转移是能够产生 He(rietta)La(cks)基因组的。但是,癌基因组与父本细胞的基因组有着很大的不同,比如 HeLa 细胞就有着 82 对染色体。Leigh Valen 不这么认为,他认为,HeLa 细胞是由同代的新品种 Helacyton gartneri, Van Valen & Maiorana 产生的。HeLa 细胞在细胞培养基中进行培养时有很多的限制因素。所有的 HeLa 细胞都是从 Lacks 的原始细胞中获取的。试管中培养的细胞实际上是 Helacyton gartneri，但是它们的原细胞会被快速生长的来自于 HeLa 的治病细胞的新细胞逐渐夺去了主导地位。

如果癌细胞是一种自由生长的种类，那么它们在人体中的出现往往意味着它们是寄生的或者共生的。HeLa 细胞是一种新物种，目前还没有发现有哪一种微生物是其寄生体或者共生体。所以，Van Valen 认为 HeLa 是一种基因共生进化的产物。这个理论认为癌细胞的不正常是一种基因共生进化的结果。在干细胞接受信号开始修复受损的细胞之前，这些癌细胞就已经存在于干细胞中了。

癌细胞没有线粒体或者它们的线粒体遭受了严重的损害，因而不能利用血液中的氧气来氧化磷酸化制造能量，这样就使得癌细胞变得相比于正常细胞是不正常的，或者说，癌细胞是从具有厌氧代谢功能的共生细胞中获取了相应的基因组。

通过基因共生的癌细胞的进化

内共生和共生起源

生命的最小单位是细胞。基因组是存在于细胞中的。细菌是一种单细胞的微生物，而植物和动物是多细胞的生物。细菌能够生存的最少的基因组是 500 个基因左右，而植物和动物具有的基因数是这个数量的 100 多倍。新达尔文主义进化理论认为，基因的变异、重组和自然选择使得基因有着很大的变化性。但是也有一小部分的生物学家不同意这样的观点。1927年，Otto Warburg 发现癌症细胞中的线粒体的功能紊乱。解剖学教授 Ivan Emmanuel Wallin 发表了《共生和物种起源》，认为原核生物通过获取共生体进化的线粒体基因组来获得他们好氧代谢机理的基因。Wallin 那时还不知道 Konstantin Merezhkovsy 或者是 Andrey Famintsin 已经做的工作，他不知道微生物共生体的获取和遗传被称为基因共生，他也不知道欧洲人其实也已经发表了同样的理论——共生导致了进化改变。

　　Lynn Margulis 基于分子生物学和古生物学的新发现，近来重新修订了获取基因组的基因共生进化理论。Kwang Joen 的实验结果显示，被一种细菌感染的阿米巴虫从那种细菌那获取了基因，从而得到了一种新的物种。Lynn Margulis 也在她的工作中描述和考虑了 Kwang Joen 的这个试验结果。

　　本来独立生长的微生物倾向于入侵其他形式的生物，并形成共生基因。一个共生细胞被另外一个包围，实现这个过程的第一步是共生关系的形成。这种内共生的关系可以用阿米巴虫的例子很好地说明。阿米巴虫这种单细胞的生物实际上类似一些共用一个能变形的细胞壁的细胞质。在这些细胞质中是各种各样的细胞体，最常见的是细胞核。有些种类只有一个细胞核，但是也有一些有很多的细胞核。除了这些细胞体外，细胞质中还有很多的其他的内含物。比如 Flurescent 细菌，是一种在大阿米巴虫 Pelomyxa palustris 的细胞质中生存的单细胞的微生物,内共生体是产甲烷细菌。所以在接种一段时间后,P. palustris 会产生甲烷气体。Pelumyxa 细胞没有线粒体,能量是通过细胞器官的糖酵解过程获取的,能量的来源是由共生的产甲烷细菌产生的甲烷气体提供的。

共生起源进化

生命大约开始于40亿年前。由细菌细胞组成的生物是没有细胞核的。它们不再分成各类,而都属于单独的一类。物种是指有核生物,它们是在25亿年前的原生代时期最早形成的。那个时候的物种是由共生起源形成的,包括细菌等微生物和大的生命体。从共生物种得到染色体组是物种起源的共生起源理论基础。现在大多数的生物学和解剖学的科学家都已经开始接受这样一个革新性的理论。

原核生物的细胞核有细胞膜包被,它们都是通过共生进化而来的。线粒体、纤毛、微管等真核生物的结构都是各自独立的,而后它们之间相互的融合通过个体核染色体的接触交换而进化出新的物种。真核生物细胞的融合产生了真菌、植物、动物以及其他种类的生物,从而组成了世间万物。

在真核生物出现之前,细菌基本上进化出现了重要的新陈代谢的所有特征。细菌的代谢方式是其对地球大气历史的一个反映。原始大气主要由二氧化碳、水和其他一些火山喷发物组成。38亿年前,厌氧细菌能去除空

气中的 CO_2,这使得细菌能够进行碳循环,使地球避免像金星那样过热烧毁的命运。消耗的 CO_2 最后可能都被转移到了火星上了,因为火星上可能还存在着一些厌氧细菌。火星上独特的气候将 CO_2 气体变成 CO_2 干冰。地球由于30亿年前进化出了产甲烷细菌,所以能够避免火星的命运。矿物质碳转化为甲烷,使得远古时代的地球能在太阳辐射能降低的时候保持温度和热量。

$$2C+2H_2O=CH_4+CO_2$$

但是,甲烷的持续产生也造成了地球的过热化。这时地球就会进行自我调节。同位素研究表明,25亿年前产甲烷细菌是不断在减少的,它们逐渐地被甲基营养型细菌和固氮蓝藻所代替。前者通过利用矿物质氧来氧化甲烷而产生能量,供细菌细胞生长发育。

$$CH_4+2NO_2^-=CO_2+2H_2O+2NO^-$$
$$2NO^-+4H^+=2H_2O+N_2$$

甲基营养型细菌进行的这种反硝化反应也将氮气转移到了空气中。另外,固氮蓝藻类还可以通过光合作用制造氧气。

地球进化历史上的氧气的生成,是25亿年前最大的变化。氧气的产生推动太古代走向原生代。固氮蓝藻类成为原生代的最主要的生物,它们通过利用 CO_2 来制造氧气。这样就减少了地球的温室效应,使得地球逐渐走向冰期和寒武纪。生命历史是从产生 CO_2 或者吸收 CO_2 的生物的出现开始的,导致了冰川时期(大约3亿年前)和地球的剧烈变暖时期(1.5亿年前,侏罗纪后期)的开始。

地球上生命出现的历史也是共生起源的历史。植物细胞中的叶绿体能够利用阳光将 CO_2 和水合成能量。实际上叶绿体是从以前独立生存的光合细菌进化来的,即植物从该种细菌中得到了基因组。线粒体能够利用氧气,将有机物分子消化代谢掉,实际上是从独立生活的类似于 bdellovibvirio 的一种细菌进化来的,即动物和植物得到了基因。在进化学历史的框架下,我们可以说动物和植物细胞线粒体的共生起源的开始定位于氧气产生的时期。氧气对于厌氧的生物来说是有毒的,氧气的出现打破了之前几乎全是厌氧的产甲烷菌占主导地位的局面,那个时候只有很少数的存在于厌氧环境中的细菌能够生存下来。植物和动物能通过线粒体的呼吸作用不断地进化。

在厌氧环境中的幸存者就需要去寻找新的共生的搭档了。阿米巴虫 Pelobiontida 是一种真核生物,但是这种真核生物却没有线粒体。那些原始的 amitochondriteate 在早期进化的原生生物不断增多后也逐渐变多了,它们是通过获取与真细菌、太古菌和真核细菌的基因相同的共生基因序列而不断进化的。Pelobiontida(泥生目)生物并没有线粒体,是因为它们在线粒体内共生开始前就开始脱离线粒体生活了。那些生物的共生体主要是一些在大气中出现氧气之前进化过来的厌氧代谢的生物。

癌细菌也像大阿米巴虫那样获取它们的代谢方式。独立生活的真核生物从 Desulforvibrio sp. 或者其他一些能够将亚硝酸盐转化为葡萄糖的厌氧细菌中获取需要的基因组。癌细菌像大阿米巴虫一样是在大气中没有氧气、植物和动物通过共生获取线粒体的时候之前就进化了。当植物和动物进化后，它们不仅与那些有线粒体的细胞发生融合，它们还获取了能在人类胚胎中制造癌性干细胞的癌的染色体。但是它们还是处于休眠状态的，只有在线粒体处于像 25 亿年前的环境中时才会被激活。这时，具有能进行厌氧代谢的细胞会占据主导地位，这样肿瘤就产生了。而肿瘤的生长是需要持续不断地提供亚硝酸盐的。在这样的理论框架下，亚硝酸盐不是一种能导致癌生长的致癌物质，而是一种维持癌细胞生长的致癌物质。当饮用水中没有了亚硝酸盐，那么癌细胞的营养供给就被切断了，癌细胞就会被"饿死"，就不会有癌细胞的生长了。

亚硝酸盐维持癌的生长

中国癌症研究中心前主任 Lu 院士曾和他的同事们一起到河南一个癌症流行的地方林县，调查癌症传染流行与饮用水中高亚硝酸盐浓度的关系。他们在那里做了很多临床试验和研究，首次发现人类食管癌是由含 N 的亚硝胺类物质引发的。他们在 1989 年报道说：

"在目前的研究中，取自林县（一个食管癌高发的地区）的人类胎儿的食管上皮细胞在 NMBzA 培养液中培养 3 个星期。然后，将这些培养的细胞异种移植到 BALB/c 小老鼠的肠系膜上，这些老鼠是用含有亚硝胺的饮用水喂养了8个月的。结果显示,接种了NMBzA的老鼠在细胞移植2个月后就被发现了其肠系膜上有小的肿瘤。8个月后,这个肿瘤长到了2.18cm。但是在对照的小老鼠中没有发现肿瘤……对肿瘤进行的病理学研究表明存在鳞片状细胞癌。从肿瘤细胞中提取的DNA与Alu序列杂交……结果表明肿瘤是与人类肿瘤一致的。"

在林县进行的试验是一个很困难的试验:学者们成功地在含亚硝基物质的培养液中培养出癌干细胞之前,经过了超过300多次的试验。癌细胞是非常少的,这个试验表明癌干细胞被激活了,300个试验中有一个试验的食管癌发生。饮用水中的亚硝酸盐对人类健康造成的影响是Lu院士亲自告诉我的,是根据以上用含有亚硝基喂养的小老鼠在接种癌组织细胞8个月后就出现癌变的试验结果Lu院士才下的这个结论。严格地来讲,癌不是由饮用水中的亚硝酸盐引起的,而应该说癌是由饮用水中的亚硝酸盐维持生长的。

正常的细胞是通过线粒体的器官从血液中得到氧从而将葡萄糖转化为能量的,我们的根本假设是细胞是从亚硝酸盐中得到氧来将葡萄糖转化为能量。就像氧气对于正常细胞的作用一样,亚硝酸盐对于癌细胞也是非常重要的。我们体内的正常细胞不会被血液中存在的亚硝酸盐破坏,它们只有在缺氧时才会死亡。亚硝酸盐本身不是毒药,血液中过量的亚硝酸盐在极端情况下会引起高铁血红蛋白症,这在临床中已有发现。但是亚硝酸盐不会引起正常细胞长癌,所以,正如欧洲和美国的专家说的那样,被试动物在摄取了亚硝酸盐后也不会引起它们生癌。

正如细胞正常生长需要氧,癌细胞也需要亚硝酸盐。癌细胞的生长是通过持续不断地供给亚硝酸盐来维持的。当饮用水中的亚硝酸盐来源被切断后,癌细胞就会"饿死",癌症就会消失了。这样的机理也可以解释为什么癌症死亡率在中国的一些饮用水中含亚硝酸盐多的地区会很高,而当他们换了饮用水的水源后就能够避免癌症的传染流行。

Lu 院士也解释了为什么吸烟会引起癌症。烟草中的亚硝酸是和烟碱一起组成有致癌作用的亚硝基类物质的。而后,当癌细胞被激活后,亚硝酸盐就能持续不断地提供给癌细胞生长所需。

癌的预防和治疗

在癌的预防和治疗方面,我们建议读者查阅"Warburg 效应",即癌是由于缺氧条件造成的,建议要多补充氧气,比如呼吸新鲜的空气进行临床治疗。我们则向读者推荐一种更便宜自然的预防和治疗方法,即饮用不含亚硝酸盐的饮用水。

Hsu-Matsumoto 理论认为癌细胞是在进化中获取相应的基因而形成的,这个结论得起考验。我们和世界各地的同事们一起对癌症死亡率和饮水中亚硝酸盐含量的关系进行了系统的分析。我们将不含亚硝酸盐的饮用水引入到"癌症村"中供人们饮用,并分析癌症死亡率会不会因此而相应降低。我们准备和癌生理专家一起合作研究体内细胞中是哪一种器官在共生进化的时候获取了Desulforvibrio 和甲基营养型细菌中获取了致癌的基因。

在中国那些饮用水中不含亚硝酸盐的地区，癌症的发生率和死亡率显著下降。我们觉得欧洲和北美洲的民众不能因为专家们专注于在学术争论中获胜而经受患癌症的痛苦。即使我们的理论被证明是错误的，至少饮用纯净水是不会有什么致病的危害吧。一个人可以不改变他的生活方式，他可以继续在早餐时吃熏肉和鸡蛋，尽管熏肉中含有大量的亚硝酸盐。如果我们的理论是正确的，癌细胞在我们体内都是存在的，但是如果没有持续供给亚硝酸盐，癌细胞是不会生长的。当没有亚硝酸盐供应时，癌细胞就会像他们"躲"在骨髓中的最初的状态一样处于休眠状态。当癌干细胞从骨髓中脱离进入血液、肺部、消化系统等这些亚硝酸盐可能存在的地方时，我们就需要注意防范了。如果我们的理论是正确的，那么癌症患者就可以通过让癌细胞"饿死"的方式杀死癌细胞，那么他就应该在确诊患癌后开始饮用不含亚硝酸盐的饮用水。

4 Role of Nitrite in Tumor Growth, Symbiogenetic Evolution of Cancer Cells, and China's Successes in the War against Cancer

Abstract

Statistics and experiments indicate a correlation between cancer mortality and nitrite iin drinking water. Nitrite is a reductant that can deprive a cell of oxygen; it is also an oxidant that can be a substrate in anaerobic ammonium oxidation, the metabolic mode of the anammox bacteria. Eukaryote cells evolved through a fusion of anaerobic and aerobic bacteria. We postulate that an anammox bacterium sought refuge in a fusion with a membrane-bound cluster of aerboic bacteria. While the latter evolved into mitochondria organelles, the former became the nucleus of a prokaryote cell. Eventually, oxidative phosphorilation is the characteristic metabolic pathway of normal eukaryote cells, and we postulate that anammox is the protein-catabolism pathway for cancer cells. The metabolism consumes nitrite and explains thus the link between nitrite and cancer.

The Nitrite Paradox

The relation between nitrite and cancer is controversial. Experimental studies evalusted by the Life Science Council of the National Academy of Sciences of the United States and by the Committee of Experts on Health Control of Foodstuffs of the Council of Europe indicated no carcinogenic effect of nitrite. Statistical studies in North America and Europe have not yielded a definitive conclusion. The EPA of the United States, followed by the WHO, identified a no-observed-adverse-effect level (NOAEL) of 10mg/L nitrate-N, and of 1mg/L nitrite-N.

Experiments by Russian, Japanese, and Chinese scientists hava found rapid proliferation of cancer cells cultured in nitrite solutions. Experiments denying a link between nitrite and cancer may hava been a misinterpretation of experimental data by scientists who made no distinction between tumor-initiation and tumor-promotion. The statistical data linking cancer to nitrite are even more impressive. After the rapid industrial development of China after 1980, there have been remarkable increases in the annual cancer-mortality rate at places where the nitrite concentration in drinking water has substantially increased. The Chinese statistics are obvious because of the great local differences in nitrite pollution. In Europe an

North America, nitrite pollution is ubiquitous and the differences are less notable.

This paper seeks an explanation of the Chinese observation of a link between nitrite and cancer.

Warburg's Effect

Gene mutations may have been caused by deletion or insertion of oncogenes. Mutations that lead to malignancy cause "immortality", allowing malignant cells to escape destruction. Eventually cancer cells break free from the tissue of which they are a part; they can grow and multiply without being attached. Malignant cancer cells travel via the bloodstream to start more tumors elsewhere. The process of metastasis is final.

The stem-cell hypothesis postulates that cancer cells are inherited. Stem cells are called to respond to signals to regenerate damaged cells. Only a very few of such are cancerous, but they are able to differentiate into other tumor cells and malignancy.

The origin of cancer is a multi-factorial process. The growth and reproduction of cancer cells depend on metabolism, by which organisms acquire and utilize the energy they need for maintenance of life and performance of biologic functions. There are two main ways to generate energy from the catabolism of glucose in human cells in the form of adenosine triphosphate (ATP): oxidative phosphorilation and glycolysis. Normal cells rely on the pathway of the former; oxygen in the blood stream is taken up by the mitochondria organelles in cells, 38 moles of ATP are derived from the oxidation of 1 mole of glucose, yielding a vast amount of energy. The metabolic mode switches to glycolysis at times of oxygen deprivation. Glycolysis is a less effective metabolism, producing only 2 ATP molecules per glucose molecule.

Warburg noted a difference between cancerous and normal cells in the ratio of glycolysis to respiration; cancer cells seem to be adapted to the hypoxic condition inside tumors. He proposed, therefore, that the prime cause of cancer is the replacement of the respiration in normal body-cells by fermentation. The Warburg hypothesis receiced support when it was noted that cancer cells with dysfunctional mitochondrial function do not respire by oxidative phosphorilation, when glycolysis becomes the energy-producer.

Glycolysis can, however, take place under aerobic conditions in cancer cells. The glycolysis mode of metabolism is an expression of dysfunctional mitochondria, and cannot be considered a cause of tumor-initiationr or tumor-promotion.

Recent studies noted that oxygen is consumed during transport across the basal membrane to produce anaerobic micro-environment in cancer cells. Warburg could be correct in his postulate that cancer cells grow under locally hypoxia conditions induced by the oxidation of nitrite. Cells with dysfunctional mitochondria normally die, when they are neither protected nor fed. Kremer suggested that the metabolic pathways in cells with dysfunctional mitochondria can be reverted to anaerobic ones. With an inspiration from Matsomoto, I started my search for the ancestry of cancer cells in anaerobic bacteria.

Anaerobic Metabolism by Anammox Bacteria

Anaerobic metabolism is the mode of methanogenic, sulfate-reduction and methylotrophic bacteria; the oxidants are carbonate, sulfate, and methane. Anaerobic ammounium oxidation by nitrite(anammox) is theoretically possible. It was, however, deemed impossible that such a chemical operation could be performed by anaerobic bacteria, until Arnoid Mulder detected the reaction in wastewater treatment. Anammox bacteria, belonging to the group Planctomycetes, were identified by Kuenen and his colleagues in sewage sludge. The overall anammox process consists of the following two denitrification reactions:

$$\text{Catabolic}: NH_4^+ + NO_1^- = N_2 + 2H_2O \qquad (1)$$

Anabolic: $CO_2 + 2NO_2^- + H_2O = CH_2O(biomass) + 2NO_3^-$ \hfill (2)

Thus, NO_2^- is not only there-acceptor in ammonium oxidation but also the electron donor for CO_2.

The prokaryote group Planctomycetes is identifiable by its S16r RNA gene. The anammox bacteria are distinctive for their budding reproduction and cell organization. Being chemo-litho-autotrophic organisms, they gain their energy(reaction(1)) and biomass for cell growth(reaction (2)) from the conversion of the inorganic substrates ammonium and nitrite. while cell carbon is derived from carbon dioxide.

In bio-reactors containing sewage sludge, oxygen is first pumped in for the nitrification of ammonium, before the anammox process starts to de-nitrify and produce lipids to build cells. Planctomycetes cells grow by anaerobic ammonium oxidation and the increase of the biomass has a doubling time of about two weeks. The biomass is decreased when the annamox process is not functioning.

Clusters of Planctomycetes cells are bound by a biomembrane. Nitrite from the oxygen-deficient water of sewage sludge diffuses through the membrane into the cells for the anammox metabolism. The presence of nitrite insures that the micro-chemical environment inside the membrane remains anaerobic. The ammonium pro-

duced from the solution of ammonia in a sewage sludge is at first converted to nitrate by the oxygen pumped into a bio-reactor:
$$NH_4^+ + 2O_2^- = 2H^+ + NO_3^- + H_2O \qquad (3)$$
before nitrate is further reduced by ammonium into nitrite(nitrification)
$$NH_4^+ + NO_3^- + O_2 = 2H^+ + 2NO_2^- + H_2O \qquad (4)$$
$$2NH_4^+ + 3O_2 = 4H^+ + 2NO_2^- + 2H_2O \qquad (5)$$

The oxidant nitrite thus produced by reactions (4) and (5) is the substrate to react with the reductant ammonium for anammox (reaction(1)).

The autotrophic bacteria exist in bioreactors of mixed cultures. Experiments indicate that anammox bacteria metabolize ammonium and nitrite anaerobically. Their cells have an internal anammoxosome compartment. The anammox ability of Planctomycetes is catalyzed by an enzyme (Hydroxylamine oxido-reductase, or HAO) encoded by the S16rRNA gene. The hydroxylamine oxidoreductase (HAO) is localized within the compartment. The anaerobic ammonium oxidation by hydroxylamine is the core of the protein catabolism. Where the catabolic reduction of NO^{2-} (reaction (1)) is not complete, the products are N_2O and NO instead of N_2. The anabolic process (reaction (2)) leads to the formation of lipids to build cells.

Nitrosococcus oceani, Urease, and Production of Ammonia

The complete genome of an anammox bacterium Nitrosococcus oceani has been deciphered. N. oceani is identiffed by its 16S rRNA sequencing. The bacterium has all the genes needed for encoding enzymes for anaerobic metabolism. It has the genes to encode two complete terminal cytochrome C oxididase to gain energy through oxidation of organic matter. It has the genes to encode AMO and PMMO to oxidize ammonia to hydroxylamine and enzyme polyphosphate kinase for glycolysis. It has genes to encode a fructokinase that can make fructose-6-phosphate which regulates glycolysis. It has the genes for the ion transport of nitrite imported via format-nitrite(FNT family)or via ammonia/ammonium permease, and for the export of organic and toxic compounds. Most important of all, N. oceani has genes to encode enzymes for the hydrolysis of urea to ammonia and CO_2, providing thus the cell with the sole source of energy and the reductant ammonium. N. oceani does not contain genes that encode subunits of hydrogenase; it cannot depend on H_2O as the sole source of energy and reductant.

An Evolutionry Theory of Cancer Origin

The anammox bacteria belong to the most ancient of life forms. The terrestrial atmosphere had no oxygen in the Archaean Era, when anaerobic ammonium-oxidation was a metabolic process that built cells and converted nitrite into the nitrogen of the atmosphere. The anammox bacteria are identifiable by their 16SrRNA gene—a gene that is identified in the genomes of all eukaryotes.

Lynn Margulis proposed that the eukaryotic cell originated from a fusion of an anaerobic and an aerobic bacterium. Kremer presented a hypothesis of metabolism-regression of cancer cells, when he postulated that the cancerous genes are inherited from the anaerobic bacterium. Lane proposed that the invader was an anarchaeon with a genome similar to that of the parasite Rickettsia prowazekii—an aerobic bacterium that used oxygen to generate its energy, and that the host should have been an anaerobic bacterium. Considering the role of nitrite in the growth of tumor ceils, I suggest that an anammox bacterium with genes to encode all the enzymes that produce nitrite and ammonium for anaerobic ammonium oxidation was the invader that effected the fusion to produce the first eukaryote cell. A cluster of aerobic bacteria bound by a bio-membrane was the host. The invader evolved into the nucleus, and the aerobic bacteria are changed into the mitochondria of eukaryote cells.

The origin of eukaryotes came after the evolution of photosynthetic bacteria. Some 2.5 billion years ago, oxygen entered into the atmosphere, so that aerobic bacteria flourished. While the global anaerobic environments were being destroyed during the Oxygen Revolution, anaerobic bacteria had to seek refuge in locally anaerobic environments. Where a cluster of aerobic bacteria, in a similar fashion as the anammox Planctomycetes cells, is enclosed in a bio-membrane, the micro-environment inside could be nearly anaerobic because of the respiration by the aerobic bacteria. The micro-hypoxic environment became thus the refuge for the anaerobic invader.

Biologic denitrifying activities were once thought to be uniquely characteristic of prokaryotes, before the discovery of the fungal denitrification systems. They are induced under the same conditions as those inducing the bacterial systems, i. e. , the presence of nitrite and a low oxygen-fugacity micro-environment that permits anaerobic ammonium oxidation. Nitrite reductase has been detected in the mitochondrial fraction from denitrifying cells. Respiratory substrates such as malate, peruvate and succinate, were effective donors of electrons to these activities in the mitochondrial fractions. Nitrite reduction is coupled to the systems to produce ATP energy.

It was discovered last year that plants also have an anaerobic respiratory system. Plant mitochondria under anaerobic conditions have a capacity to use nitrite as an electron acceptor to generate ATP. Mitochondria isolated from the roots of barley and rice seedlings are capable of oxidizing external NADH and NADPH anaerobically in the presence of nitrite. The rates of anaerobic ATP synthesis are of the same order of magnitude as glycolytic ATP production during anoxia, but are only about 3%-5% of the aerobic mitochondrial ATP synthesi srate.

Adoption of anaerobic ammonium oxidation by animals is yet to be proven. Considering that anammox of eukaryotes was unknown until a few years ago, the possibility is not excluded that this mode of metabolism has been adopted by cancer cells. Like the Planctomycetes cells enclosed in a bio-membrane, cancer cells are also separated by a basement membrane from blood vessels, which are conlined to the stromal compartment. Substrates, such as oxygen, nitrites, etc., must diffuse across the membrane. The existence of a O_2 gradient has been detected, and the micro-environment is anaerobic near the oxygen diffusion-limit some 200 μm from a vessel wall. There is also a pH gradient, with pH decreases from 7.3 to 6.7 in a 400 μm distance. The pH decrease is an indication that the anammox process has become effective in the anaerobic micro-environment of cancer cells, where ammonium is oxidized by nitrite.

It is well known that humans excrete more nitrites and nitrate than they could have ingested from food. The presence of intra-cellular nitrites and nitrates, the metabolic products of anammox(reactions(4)and(5)) could be considered an evidence of anammox metabolism by cancer cells. The other substrate ammonium for anammox is provided by the hydrolysis of urea by the enzyme urease; the reaction produces ammonia and CO_2:

$$(NH_2NH_2)_2CO + H_2O = CO_2 + 2NH_3$$

The enzyme urease that promotes the hydrolysis of urea is present both in the genomes of anaerobic bacteria such as N. oceani and in cancer patients.

Multi-purpose genes in the human genome are over expressed in cancers. Of these genes, 15 are involved in the protein biosynthesis and folding, six in glycolysis. A comparison of the human cancer genomes with that of N. oceani may verify or falsify whether the anammox processes are the metabolic pathways of cancer cells.

In cancer cells, genes for oxidative phosphorilation is deleted or suppressed, replaced by those that encode glycolysis. Even more significant is the observation of the enhanced activities of hydrolysis of urea by the enzyme u rease in the metabolism of cancer patients. In our postulate that anammo promotes the tumor growth, ammonium is produced by the hydrolysis of urea for nitrification. This ammonium is then combined with nitrite as the substrates for anammox mode metabolism of cancer cells.

Prevention and Cure of Cancer

Current cancer researches in the West emphasize the studies of tumor-initiations. The mechanism of gene mutations are unpredictable. The treatments, including surgery, radiation and chemical therapies, remove or exterminate cancer cells. Some cancer drugs or therapies have, however, a function to suppress enzymes such as HAO, AMO, urease, etc that promote cancer-growth.

My hypothesis of symbiogenetic origin of cancer is a chain ot facts and postulates linked by speculations: cancer mortality correlates to nitrite in drinking water—nitrite is a tumor promoter—glycolysis is prevalent in cancer cells—anaerobic respiration is the metabolic mode of cancer cells—ammonium and nitrite are the substrates for anammox under hypoxic conditions—anammox is the metabolic mode of cancer cells—urease promotes the production of ammonium, which is oxidized in part (during anammox) to form nitrite—suppression of enzymes that produce ammonium and/or nitrite could prevent anammox and thus starve the cancer cells. Cancer research studies could falsify or verify the postulates or speculations to yield new drugs for cancer cure.

The main purpose of this paper is, however, to provide an explanation of the Chinese observation of a link between nitrite and cancer. We believe that China's victorious battles in the War Against Cancer can be explained because their recognition of a link between nitrite and cancer-mortality.

It has been suggested that daily micrograms of nitrite-intake are insufficient to provide the substrate for protein-catabolism to grow cancer cells. However, the role of nitrite in carcinogenesis is probably that of a catalyst to induce hypoxia that triggers the hydrolysis of urea and the ammonium oxidation. It is well known that nitrite in the bloodstream could induce a loss of oxygen, and excessive dose is the cause of methemoglobinemia. C. V. Dang suggested that the role of nitrite-intake is to induce hypoxia. I now believe that the large quantity of nitrite needed for anammox metabolism may have been produced endogenously by the hydrolysis of urea.

Taking seriously the implications of their statistical data, the Chinese Ministry of Health promulgated in 2004 a maximum permissible contamination level (MPCL) of 0.002 mg/L nitrite nitrogen for "purified clean water", —

1/500th the NOEAL of the US or of WHO. The Chinese Government has spent billions to switch the sources of drinking water in areas of cancer epidemics: in Henan, Guangxi, Fujien, Tienjien, Jiangsu, etc. The policy has yielded results, and China could claim local victories in the War against Cancer, where the cancer mortality rate has been reduced to half after the switch of the sources of drinking water (see Tables 1 and 2). The Chinese Government is appropriating, for example, 1.6 billion yuan to provide nitrite-free groundwater for the 3.8 inhabitants of Zhoukou, Henan. To maintain balance of hydrologic budget, the Nation is embarking on a multibillion $ program of denitritization, using my invention of biologic cleansing by diatoms. We expect to prevent more than half a million cancer deaths in China each year, after the Nation's sources of water supply meet the 0.002 mg/L nitrite-N standard recommended by the Chinese Ministry of Health.

Acknowledgements

We are grateful to James Wang, Nick Lane, David Roger, Graham Shields, Lynn Margulis, Frank Dickson, and Zeng Yi for critical reading of the manuscript. The research is supported by the Center of Health and Environmental Engineering, Henan Unicersity, Kaifeng China.

References Ciled

1. National Academy of Sciences, National Research Council, Life Sciences Council, "The Health effect of Nitrate, Nitrite and N-Nitroso", National Academy Press. Washington DC. 1981.
2. Committee of Experts on Health Control of Foodstuffs, "Nitrates and Nitrites in Foodstuffs", Council of Europe Press, Strasbourg. 1993. pp. 128.
3. D. C. Paik, D. V. Saborio, R. Oropeza and H. P. Freeman, "The Epidemiological Enigma of Gastric Cancer Rates in the US", Internationgal Journal of Eidemiology, Vol. 30, No. 1, 2001. pp. 181-182. doi:10. 1093/ije/30. 1. 181
4. Y. Kitamura, T. Umemura, K. Okazaki, et al. , "Enhancing Effects of Simultaneous Treatment with Sodium Nitrite on 2-Amino-3-methylimidazo [4,5-f] Quinoline-Induced Rat Liver, Colon and Zymbal's Gland Carcinogenesis after Initiation with Diethylnitrosanline and 1, 2Dimethylhydrazine", Internationa Journal of Cancer. Vol. 118, No. 10, 2006. pp. 2399-2404. doi:10. 1002/ije/21649
5. T. H. Tseng, M. C. Chang, J. D. Hsu, et al. , "Tumor Promoting Effect of N-nitroso-N-(2-hexanonyl)-3-nitrotyramine (a Nitrosated Maillard Reaction Product) in Benzoa Pyrene-Initiated Mouse Skin Carcinogenesis", ChemicoBiological Interactions, Vol. 115, No. 1, 1998. pp. 23-38. doi: 10. 1016/s0009-2797(98)00056-8
6. H. Tsuda and M. Hasegawa, "Malignant Transformation of Mouse BALB/c3T3 Cells Induced by $NaNO_2$", Carcinogene-

sis, Vol. 11, No. 4, 1990. pp. 595-597. doi:10. 1093/carcin/11. 4. 595
7. J. Sun, K. Aoki, W. Wang, et al. , "Sodium Nitrite-Induced Cytotoxicity in Cultured Human Gastric Epithelial Cells", Toxicology in Vitto, Vol. 20, 2006. pp. 1133-1138. doi:10. 1016/j. tiv. 2006. 02. 005
8. O. I. Bol'shakova, A. G. Sverdlov and S. I. Timoshenko, "Effect of Nitric Oxide Donor on the Stability of NonTransformed and Malignant Cells to Ultraviolet and Gamma Radiation", Tsitologia, Vol. 46, 2004. pp. 39-42.
9. I. V. Kondakova, G. V. Zagrebel'naya and E. T. Choinzonov, "NO-Generating Compounds Modify Tumoritoxic Effect of Doxorubicin", Bulletin of Experimental Biology and Medicine, Voi. 137. No. 6, 2004. pp. 585-587. doi:10. 1023/B:BEBM. 0000042719. 10494. 3f
10. A. Yoshida, T. Harada and T. Kitazawa, "Effects of Age on Endometrial Carcinogenesis Induced by Concurrent Oral Administration of Ethylenethiourea and Sodium Nitrite in Mice", Experimental and Toxicologic Pathology, Vol. 48, No. 4, 1996. pp. 289-298. doi:10. 1016/S0940-2993(96)80021-2
11. C. S. Huangfu, K. J. Hsu, M. Z. Qin and T. S. Ma, "The Relationship between Nitrite and Cancer", Journal of Henan University(Natural Science), Volume for 2009. pp. 1-5.
12. D. Lai, "Geostatistical Analysis of Chinese Cancer Motality", Journal of Data Science, Vol. 2, 2004. pp. 177-193.
13. Z. X. Xu, "Chemical Fertilizers, Sewage-Treatment and Aesophageal Cancer (in Chinese)", Scientific Publishers. Bei-

jing, 2003. pp. 257.

14. W. Q. Lu, C. L. Chen and W. Z. Li, "Link between Cancer and Drinking Water Quality in Areas of High and of Low Cancer-Incidence", Cancel in China, Vol. 9, 2000. pp. 227.

15. J. Y. Han, M. X. Wang and Y. Ma, "Cancer Mortality Rate and Nitrite in Drinking Water, Linzhou (Chinese with English Abstract)", Henan Disease Control Medicine, Vol. 11, 2000. pp. 65-69.

16. X. Deng, C. Q. Lin and Z. G. Gu, "Cancer-Mortality Rate of Nitrite in Canton's Drinking Water(Chinese with English Abstract)", Physiological Science, Vol. 23, 2004. pp. 38-41.

17. J. Han, "The Effect of Switching the Source of Drinking Water on the Cancer the Source of Drinking Water on the Cancer-Mortality Rate of Linzhou", Journal of Envitonment and Health(Beijing), Vol. 22, 2005. pp. 200-202.

18. O. Warburg, "On the Origin of Cancer Cells", Science, Vol. 123, No. 3191, 1956. pp. 309-314. doi: 10. 1126/science. 123. 3191. 309

19. K. Garber, "Energy Boost: The Warburg Effect Returns in a New Theory of Cancer", Journal of National Cancer Institute, Vol. 94, 2004. pp. 1805-1806. doi: 10. 1093/jnci/96. 24. 1805

20. R. A. Gatenby and R. J. Gillies, "Why do Cancers Have High Aerobic Glycolysis," Nature Reviews Cancer, Vol. 4. 2004. pp. 891-899. doi: 10. 1038/nrc1478

21. H. Kremer, "The Silent Revolution in Cancer and AIDS Medicine", Xlibris Corporation, Zurich, Vol. 633, 2008. pp. 71.

22. K. J. Hsu, P. Matsumoto, W. H. Ye, Y. Zhen, S. X. Lu, J. L. Li and S. Sun, "Nitrite in Drinking Water and Cancer", Scientia Sinica, 2009, In Press.
23. A. Mulder, "Anoxic Ammonium Oxidation", US Patent No. 427849, 1992.
24. A. Mulder, A. A. Graaf, L. Robertson and J. G. Kuenen, "Anaerobic Ammonium Oxidation Discovered in a Deni-trifying Fluidized Bed Reactor", FEMS Microbiology Ecology. Vol. 16, No. 3, 1995. pp. 177-184. doi: 10. 1111/j. 1574-6941. 1995. tb00281. x
25. J. G. Kuenen and M. S. M. Jetten, "Extraordinary Anaerobic Ammonium Oxidizing Bacteria", American Society for Microbiology News, Vol. 67, No. 9, 2001. pp. 12.
26. K. Egli, J. R. vall der Meer and H. Siegrist, "The Anammox Process for Nitrogen Removal from Waste Water", EAWAG News. Zurich, Vol. 56, 2005, pp. 20-21.
27. L. van Niftrik, W. J. Geerts, E. G. van Humbel, R. J. Webb, J. A. Fuerst and M. Strous, "Linking Ultrastructure and Function in Four Genera of Anaerobic-Ammonium Oxidation Bacteria", Journal of Bacteriology, Vol. 190. No. 2, 2008. pp. 708-717. doi: 10. 1128/JB. 01449-07
28. M. G. Klotz, P. S. G. Chain, A. F. El Sheikh, L. R. Hauser, S. G. Hommes, F. W. Lariomer, S. A. Malfatti, J. M. Norton, A. T. Poret-Peterson, L. M. Vergez and B. B. Ward, "Complete Oenome Sequence of Marine Chemolithoautotrphic Amm-OX Bacterium Nitrosococcus oceani", Applied and Environmental Micribiology, Vol. 72, 2006. pp. 6299-6315. doi: 10. 1128/AEM. 00463-06

29. Sagan and L. Margulis, "On the Origin of Mitosing Cells", Journal of Theoretical Biology, Vol. 14, 1967. pp. 255-274.
30. N. Lane, "Power, Sex, Suicide: Mitochondria and the Meaning of Life", Oxford University Press, Oxford, 2005. pp. 354.
31. M. Kobayashi, Y. Matsuo, A. Takimoto, S. Suzuki. F. Maruo and H. Shoun, "De-Nitrification, a Novel Type of Respiratory Metabolism in Fungal Mitochondrion", Journal of Chemical Biology, Vol. 271, No. 27, 1996. pp. 16263-16267. doi: 10. 1074/jbc. 271. 27. 16263
32. M. Stoimenova, A. Igamberdiev, K. J. Gupta and R. D. Hill, "Nitrite-Driven Anaerobic ATP Synthesis in Barley and Rice Root Mitochondria", Planta, Vol. 226, No. 2, 2007. pp. 465-474. doi: 10. 1007/s00425-007-0496-0
33. H. H. Mitchell, H. Shoule and H. S. Grindley. "The Origin of Nitrates in Urine", Journal of Biological Chemistry, Vol, 24. 1916. pp. 461-490
34. J. O. Lundberg, J. Weitzberg, J. A. Cole and N. Benjamine, "Nitrate, Bacteria and Human Health", Nature Revieuw/Microbiology, Vol. 2, 2004. pp. 593-602.
35. S. Ito. T. Kohli, H. Murakita, Y. Ohotaki, T. Azuma and M. Kruiyama, "Differences in Urease Activity in the Live Helicobacter pylori Cultured from Patients with Gast groduodenal Diseases", European Journal of Gastroenterology and Hepatology, Vol. 7, 1989. pp. 83-88.
36. B. Alternberg, C. Gemuend and K. O. Greulich, "Ubiquitous Cancer Genes: Multipurpose Molecules for Protein Micro-Arrays", Proteomics, Vol. 6, 2007. pp. 67-71. doi: 10. 1002/pmic. 200500154

37. C. H. Wu, M. Huang, C. S. Yeh, J. W. Wang, T. L. Cheng and S. R. Lin, "Over Expression of Helicobacter Pylori-Associated Urease mRNAs in Human Gastric Cancer", DNA and Cell Biology, Vol. 2, 2007. pp. 641-648. doi: 10. 1089/dna. 2007.0599
38. J. Wang, Harvard University, June 2007.
39. C. V. Dang, D. Dean and J. Hopkins, Medical School, June 2007.
40. K. J. Hsu, W. H. Ye, Y. H. Kong, D. Li and F. Hu, "Use of Hydrotransistor and Self-Cleaning Pond for De-Nitrification to Produce Purified Non-Carcinogenic Water for Drinking and for Groundwater Recharge", Technical Report, Tarim Resources Recycling Ltd., Haslemere. 2007.

亚硝酸盐在肿瘤生长中的作用,兼论癌细胞的共生进化起源与中国抗癌的成就

摘 要

统计和实验表明,癌症死亡率和饮用水中的亚硝酸盐存在一定的相关性。亚硝酸盐是一种还原剂,它夺取细胞中的氧;同时,它也是一种氧化剂,是厌氧脱氮细菌新陈代谢作用进行氨氧化的底物。真核细胞的起源来自于好氧和厌氧细菌融合并不断进化,推测原因可能是厌氧氨氧化(厌氧脱氮)细菌与好氧细菌群在彼此融合的膜中寻求避难,当后者演变成线粒体细胞器,前者便成为一个原核生物细胞的核。最终,氧化磷酸化是正常的真核细胞特有的能量代谢途径,由此我们推测厌氧氨氧化是癌细胞蛋白质分解的途径;新陈代谢消耗了亚硝酸盐,癌症的发展和亚硝酸盐之间存在关联。

亚硝酸盐悖论

亚硝酸盐与癌症之间的关系是有争议的。由美国国家科学院生命科学理事会欧洲委员会的食品卫生管理专家委员会评估的实验研究表明:亚硝酸盐没有致癌作用。

北美洲和欧洲的统计研究中也没有明确的结论。美国环保局、世界卫生组织,确定一个饮水中不明显、不良作用水平(无害作用剂量)的健康标准:硝酸盐氮不超过 10 mg/L,亚硝酸盐氮不超过 1 mg/L。

俄罗斯、日本和中国科学家的实验发现了亚硝酸盐可以加速致癌剂导致癌的发生,并促进培养的癌细胞生长。但是,科学实验否认了亚硝酸盐与癌症之间的联系,可能是被实验数据误解了。这些科学家没有区别肿瘤启动和肿瘤促进的不同。有关亚硝酸盐与癌症的流行病学统计数据令人印象深刻。1980 年后,由于中国工业的快速发展,已经出现了饮用水中亚硝酸盐浓度大幅增加的地方,这些地方每年的癌症死亡率也明显增加。亚硝酸盐污染的地方差异很大,中国统计数据是显而易见。在欧洲和北美洲,亚硝酸盐污染无处不在,差异显著。本文意在通过对中国的调查来说明亚硝酸盐与癌症之间的联系。

Warburg 效应

癌基因的插入或者缺失可能会造成细胞基因的突变,从而导致恶性细胞逃脱机体破坏而"不死",最后,癌细胞脱离它们,形成其中一部分的组织,它们可以不受身体约束而生长繁殖。恶性肿瘤细胞通过淋巴或血液流动,向其他部位扩散,造成肿瘤细胞的转移或扩散。

干细胞假说推测,癌症是遗传的。干细胞被称为受损的细胞再生的来源,只有极少数癌变,但它们能够分化成其他肿瘤细胞和恶性肿瘤。

　　癌症的起源是一个多因素共同机体的过程。癌细胞的生长和繁殖依赖于代谢,通过代谢生物体获取它们需要的能源来维持生命,从而生物功能得以发挥。在人体细胞中葡萄糖代谢产生三磷酸腺苷(ATP)的主要途径有两个:氧化磷酸化和糖酵解。正常细胞依靠前者的途径代谢,在血液中氧气被线粒体这种细胞器所携带,1摩尔的葡萄糖氧化有38摩尔的ATP产生,这是一个巨大的能源。细胞缺氧时,糖酵解模式开启,糖酵解是一种不太有效的代谢,每一个糖分子只产生两个三磷酸腺苷(ATP)分子。

　　Warburg指出了糖酵解呼吸比例在癌细胞和正常细胞之间的差别。癌细胞似乎可以适应肿瘤组织中的缺氧状况,因此,他提出癌症的首要原因是正常的体细胞呼吸被糖酵解替换。有人指出,当癌细胞线粒体功能失调,没有进行氧化磷酸化呼吸时,糖酵解成为能源的生产者,Warburg的假说因此得到了支持。糖酵解代谢模式是线粒体功能失调的表现,但是它并不能被认为是肿瘤启动或肿瘤促进的原因。最近的研究指出,氧气透过整个基底膜向组织的运输过程中不断被消耗,远离血管的癌细胞周围氧气减少,这就在癌组织内产生了缺氧微环境。Warburg学说指出,细胞在缺氧条件下选择糖酵解。亚硝酸盐作用于线粒体呼吸,阻断氧的利用,启动糖酵解,以此使细胞耐受缺氧存活下来。亚硝酸盐以这种方式诱导癌细胞的增长、演化。线粒体功能失调的细胞,当它们不被保护或者供给营养时,通常会死亡。克雷默认为,线粒体功能失调的细胞通过代谢途径可还原底物造成厌氧环境。我们有理由相信厌氧细菌代谢、生长的一些行为与癌细胞有相同点,所以用厌氧细菌代谢理论对癌细胞代谢进行研究。

厌氧氨氧化菌的厌氧代谢

细菌厌氧代谢是产生甲烷、硫酸盐还原和甲醇的模式;氧化剂是碳酸盐、硫酸盐和甲烷。厌氧氨化合物被硝酸盐氧化在理论上是可能的。然而,这样的由厌氧菌来进行的化学反应被视为不可能,直到阿诺德穆德检测废水处理中的反应,Kuenen 和他的同事们发现,在污水污泥的厌氧氨氧化细菌,确定为腐霉菌。整体的厌氧氨氧化过程为以下两个反硝化反应:

代谢:$NH_4 + NO_2 = N_2 + 2H_2O$ (1)

同化:$CO_2 + 2NO_2 + H_2O = CH_2O(biomass) + 2NO_3$ (2)

因此,二氧化氮在氨氧化时不仅是受体,也是二氧化碳的电子供体。原核生物腐霉菌是靠它本身的 16 位核糖核酸基因识别。厌氧氨氧化合菌以其出芽方式繁殖,这和细胞增殖显然是有区别的。机体通过化学反应,从无机氨和亚硝酸盐的生物转化中获得了(反应 1)能量,以此使生物体细胞获得生长所需的能量,而细胞中的碳是从二氧化碳中获得的。

含污水污泥的生物反应器,氧气首先用于铵的硝化,厌氧氨氧化过程开始前脱硝化和生产构建细胞的脂质。Planctomycetes 细胞生长厌氧氨氧化和生物量的增加时间约为两个星期。厌氧氨氧化过程无法正常运作时生物量下降。Planctomycetes 细胞群是由生物膜的约束，从污水污泥扩散通过膜进入厌氧氨氧化代谢的细胞，在缺氧时产生亚硝酸盐。亚硝酸盐的存在确保里面的膜的微化学环境仍然缺氧。从污水污泥中除氨的一个解决方案所产生的氨是先转换成生物反应器，再泵送氧以氧化硝酸盐：

$$NH_4 + 2O_2 = 2H + NO_3 + H_2O \qquad (3)$$

(3)中的硝酸盐由于被氨转化为亚硝酸盐而进一步减少(硝化)。

$$NH_4 + NO_3 + O_2 = 2H + 2NO_2 + H_2O \qquad (4)$$

$$2NH_4 + 3O_2 = 4H + 2NO_2 + 2H_2O \qquad (5)$$

氧化亚硝酸盐由反应(4)和(5)而产生，是基质与厌氧氨氧化的还原剂，自养菌存在于混合培养的生物反应器。实验结果表明，厌氧氨氧化菌代谢厌氧氨和亚硝酸盐。细胞内部有 anammoxosome 间隔。(羟胺氧化还原酶)S16rRNA 基因编码是一种酶催化 Planctomycetes 厌氧氨氧化能力。厌氧氨氧化能力的 Planctomycetes 是 S16 rRNA 基因编码的酶催化(羟胺氧化还原酶，或郝)。羟胺氧化还原酶(郝)是蛋白质分解的核心。二氧化氮分解减少(反应 1)是不完整的，N_2O 和 NO 代替氮气。同化的过程(反应 2)导致血脂的形成。

Nitrosococcus oceani、脲酶和生产的氨

Nitrosococcus oceani 是一种厌氧氨氧化细菌,其完整的基因组已被破译。

N. oceani 的识别依其 16S rRNA 的序列进行。这种细菌有所有编码厌氧代谢所需酶的基因。它编码两个完整的终端细胞色素 Coxididase 的基因,使得细菌能通过氧化有机物获得能量。它有编码 AMO 和 PMMO 的基因,而 AMO 和 PMMO 可以氧化羟胺和多聚磷酸盐激酶中的氧,为糖酵解做准备。它有编码果糖激酶的基因,可以产出调节糖酵解的磷酸果糖 6。它有控制离子传输物质的基因,使得亚硝酸盐通过亚硝酸盐格式(FNF 家庭)或氨/铵通透酶进入细胞,而排出有机物和有毒物质。最重要的是,N. oceani 有编码水解尿素酶的基因,在这个过程中,尿素水解为氨和 CO_2 并提供了细胞所需的全部能量和还原剂铵。N. oceani 不包含编码氢化酶亚单位的基因,它不能把水作为唯一的能源和还原剂。

癌症起源的进化理论

厌氧氨氧化细菌,属于最古老的生命形式。在太古时代,地球大气层没有氧气时,厌氧氨氧化在内置的细胞代谢过程中,将大气中的氮转化为亚硝酸盐。厌氧氨氧化细菌被其 16S rRNA 基因所识别,即在对在所有真核生物的基因组发现的基因。

马古利斯琳建议,真核细胞的起源是厌氧和好氧菌的融合。克雷默提出了癌细胞的新陈代谢回归的假设,他推测癌细胞的基因是由厌氧细菌继承的。琳建议,入侵者与寄生虫立克次氏体,基因组类似古菌好氧菌,使用氧气生成能量的有氧细菌,主体应有一种厌氧细菌。鉴于亚硝酸盐在肿瘤细胞生长中的作用,我认为癌细胞是由厌氧氨氧化细菌的基因编码的酶产生亚硝酸盐和厌氧氨氧化铵的影响融合产生的第一个真核生物细胞侵略者。一个由生物膜的约束好氧菌的集群主体演变成核,而好氧菌侵略者变成了真核细胞的线粒体。真核生物起源于光合细菌的进化。大约250亿年前,氧气进入到大气中,使好氧菌蓬勃发展。虽然氧气革命期间摧毁了全球的厌氧环境,厌氧菌在局部厌氧环境中可寻求避难。好氧细菌群,有与 ANAMMOX Planctomycetes 细胞类似的方式,在生物膜内的微环境内可以因为呼吸接近厌氧好氧菌。因此,微缺氧的环境成为了厌氧菌的避难所。

在发现真菌的反硝化系统之前生物反硝化的活动一度被认为是唯一的原核生物的特点。它们处在相同的细菌诱导条件下,即亚硝酸盐的存在和低的氧密度的微环境,使厌氧氨氧化进行。亚硝酸盐还原酶已检测到在反硝化细胞的线粒体内。细胞呼吸系统分子,像乳酸、琥珀酸和苹果酸是活动中的线粒体组分的有效底物。亚硝酸盐减少呼吸链电子耦合系统,使细胞产生 ATP 获得能量。

在去年,发现植物也有厌氧呼吸系统。在厌氧条件下植物线粒体有能力使用亚硝酸盐作为电子受体,生成ATP。从大麦和水稻幼苗根部中分离出线粒体是能够氧化外部的 NADH 和 NADPH 厌氧氧化亚硝酸盐而存在的。厌氧 ATP 合成速率是和糖酵解 ATP 在缺氧过程中合成的顺序相同,但只有约 3%~5% 的有氧线粒体 ATP 合成。通过动物的厌氧氨氧化尚待证实。考虑到真核生物 ANAMMOX 的不明,直到几年前其可能性仍不排除,这种新陈代谢的模式已被癌细胞所采用。像 Planctomycetes 细胞在生物膜内一样,癌细胞也局限于基质内,由血管、基底膜隔开。如底物、氧、亚硝酸盐等,必须扩散穿过细胞膜。已检测到一个氧分压梯度的存在,从血管壁附近氧气扩散约 200 微米的微环境是缺氧环境。还有一个 pH 梯度,在 400 微米的距离,随着 pH 从 7.3 下降到 6.7,厌氧氨氧化过程已成为有效的肿瘤细胞缺氧微环境,这是由亚硝酸盐氧化铵造成的。

众所周知,人体分泌的亚硝酸盐和硝酸盐比他们从食物中摄取的还要多。细胞内的亚硝酸盐和硝酸盐,厌氧氨氧化(反应 4 和 5)的代谢产物的存在,可以作为考虑癌细胞进行厌氧氨氧化代谢的证据。其他铵的尿素酶水解反应生成氨和二氧化碳:

$(NH_2 NH_2)_2CO + H_2O = CO_2 + 2NH_3$

促进尿素水解的尿素酶基因,目前存在于厌氧细菌如 N. oceani 的基因组和癌症患者中。人类基因组多用途基因在癌症中过度表达。

这些基因,参与蛋白质合成、折叠和糖酵解。一个比较可能的是人类癌基因组参与 N. oceani 验证的厌氧氨氧化过程,后者是癌细胞的代谢途径。在肿瘤细胞中,氧化磷酸化的基因被删除或抑制,取而代之的是那些编码糖酵解的基因。更重要的是水解尿素的尿素酶在癌症患者体内其代谢增强。我们假设 ANAMMOX 可促进肿瘤的生长,氨是生产尿素水解硝化而来的。铵与亚硝酸盐 ANAMMOX 模式是癌细胞代谢的基本形式。

癌的预防和治疗

在西方,目前的癌症研究,强调肿瘤的起源。基因突变的机制是不可预知的。治疗措施包括手术、放疗和化学疗法,转移或剿灭癌细胞。有一些抗癌的药物或疗法包括抑制酶脲酶功能,能阻止癌细胞的生长。

我关于癌症起源的假说是基于一连串事实和推测。癌症死亡率与饮用水中的硝酸盐、亚硝酸盐有关。亚硝酸盐是一种癌的促进剂。糖酵解在癌细胞中普遍存在。癌细胞无氧呼吸代谢铵和亚硝酸盐的模式,低底物 ANAMMOX——厌氧氨氧化是癌细胞的代谢模式,尿素促进 producti 铵,这是部分氧化(ANAMMOX),形成亚硝酸盐。抑制生产铵和/或亚硝酸盐的酶,可以防止厌氧氨氧化,从而"饿死"癌细胞。对可以证明的癌症研究和验证的假设,可以生产新的药物来治愈癌症。

然而,本文的主要目的是,在亚硝酸盐与癌症之间的联系方面,中国人给出了许多案例。我们相信,在中国对抗癌症成功战役中是可以解释的,因为他们认识到亚硝酸盐和癌症的死亡率之间的联系。

  有人曾提出，亚硝酸盐每天的摄入量不足以提供底物用以癌细胞增长的蛋白质进行分解代谢。然而，对于癌变组织，亚硝酸盐的作用可能是一种催化剂，促使缺氧触发尿素水解和氨氧化。众所周知，血液中的过多剂量的亚硝酸盐是诱发血液高铁血红蛋白血症的原因。C. V. Dang认为摄入的亚硝酸盐是诱发细胞缺氧的原因。我现在相信细胞大多厌氧氨氧化代谢所需要的亚硝酸盐可能是由尿素水解内源性氨产生的。

  中国卫生部认真对待统计数据的影响，2004年颁布"净化的水"中亚硝酸盐氮最大允许污染水平为0.002毫克/升(MPCL)，相当于美国或世界卫生组织NOEAL的1/500。"中国政府已花费数十亿美元关闭河南、广西、天津、江苏等癌症高发区域的饮用水源。饮用水源关闭后癌症患者的死亡率已减少一半。这项政策已经取得了成果，中国可能会在抗击癌症的战争中取得局部胜利。例如，中国政府下拨1.6亿元人民币，在河南周口，为380万居民提供无亚硝酸盐的地下水。

  我们期望，在中国卫生部建议实施的供应0.002毫克/升亚硝酸盐氮的标准后，每年可以减少超过50万中国癌症患者的死亡。